Early praise for *Good Math*

Mark Chu-Carroll is one of the premiere math bloggers in the world, able to guide readers through complicated concepts with delightful casualness. In *Good Math* he brings that same skill to a book-length journey through math, from the basic notion of numbers through recent developments in computer programming. If you have ever been curious about the golden ratio or Turing machines or why pi never runs out of numbers, this is the book for you.

➤ **Carl Zimmer**
author of "Matter," a weekly column about science in *The New York Times* (http://bit.ly/NYTZimmer); and "The Loom," a *National Geographic Magazine* blog (http://phenomena.nationalgeographic.com/blog/the-loom)

Fans of Mark Chu-Carroll's lively and informative blog, Good Math/Bad Math, will find much to savor in this mathematical guide for the "geekerati." Chu-Carroll covers it all, from the basics of natural, irrational, and imaginary numbers and the golden ratio to Cantor sets, group theory, logic, proofs, programming, and Turing machines. His love for his subject shines through every page. He'll help you love it, too.

➤ **Jennifer Ouellette**
author of *The Calculus Diaries*

Good Math

A Geek's Guide to the Beauty of Numbers, Logic, and Computation

Mark C. Chu-Carroll

The Pragmatic Bookshelf
Dallas, Texas • Raleigh, North Carolina

Our Pragmatic courses, workshops, and other products can help you and your team create better software and have more fun. For more information, as well as the latest Pragmatic titles, please visit us at *http://pragprog.com*.

The team that produced this book includes:

John Osborn (editor)
Candace Cunningham and Molly McBeath (copyeditor)
David J Kelly (typesetter)
Janet Furlow (producer)
Juliet Benda (rights)
Ellie Callahan (support)

Printed in the United States of America.
ISBN-13: 978-1-937785-33-8
Printed on acid-free paper.
Book version: P1.0—July 2013

This book is dedicated to the memory of my father, Irving Carroll (zt"l). He set me on the road to becoming a math geek, which is why this book exists. More importantly, he showed me, by example, how to be a mensch: by living honestly, with compassion, humor, integrity, and hard work.

Contents

Part III — Writing Numbers

Part IV — Logic

Part V — Sets

Part VI — Mechanical Math

Preface

Where'd This Book Come From?

Growing up, some of my earliest memories of my father involve math. My dad was a physicist who worked for RCA doing semiconductor manufacturing, so his job involved a lot of math. Sometimes he'd come home with some unfinished work to do over the weekend. He'd be sitting in the living room of our house, with a scattering of papers around him and his trusty slide rule by his side.

Being a geeky kid, I thought the stuff he was doing looked cool, and I'd ask him about it. When I did, he always stopped what he was doing and explained it to me. He was a fantastic teacher, and I learned so much about math from him. He taught me the basics of bell curves, standard deviations, and linear regression when I was in third grade! Until I got to college, I never actually learned anything in math class at school because my dad had always taught it to me long before we got to it in the classroom.

He did much more than just explain things to me. He taught me how to teach. He always told me that until you could explain something to someone else, you didn't really understand it yourself. So he'd make me explain things back to him as though he didn't know them.

Those times with my dad were the foundation of my love of math, a love that's lasted through the decades.

Back in 2006 or so, I started reading science blogs. I thought that these blog things were really fascinating and really exciting. But I didn't think that I had anything to say that would interest anyone else. So I just read what others wrote, and sometimes I commented.

And then one day I was reading a blog called *Respectful Insolence,* written under the pseudonym "Orac," by a guy who was a professional cancer surgeon. He was talking about a paper written by a couple of crackpots who had drawn ridiculous conclusions from data published in a public database. Orac dismantled their arguments meticulously, explaining why the authors' claims about basic medicine and biology were ridiculous. But in reading the original paper, what struck me was that refuting the authors' misunderstanding of biology was unnecessary; their entire argument turned on interpreting graph data in a way that was completely bogus. That's when I realized that while tons of biologists, doctors, neurologists, physiologists, and physicists were blogging about their specialties, no one was blogging about math!

So I went to Blogger and created a blog. I wrote up my critique of the sloppy math in the paper and sent a link to Orac. I figured that I'd probably get a couple of dozen people to read it and that I'd probably give up on it after a couple of weeks.

But once I'd published that first post on my new blog, I thought about my dad. He was the kind of guy who wouldn't approve of spending time making fun of people. Doing that once in a while was fine, but making an entire hobby out of it? Not something he'd be proud of.

Remembering how he taught me, I started writing about the kind of math I loved, trying to help other people see why it was so beautiful, so fun, and so fascinating. The result was my blog, *Good Math/Bad Math.* It's been almost seven years since I started writing it, and my posts now number in the thousands!

When I started my blog, I thought that no one would be interested in what I had to say. I thought that I'd probably be read by a couple dozen people, and I'd give up in disgust after a couple of weeks. Instead, years later, I've acquired thousands of fans who read every post I write.

This book is my way of reaching out to a wider audience. Math *is* fun and beautiful and fascinating. I want to share that fun, beauty, and fascination with you. In this book, you'll find the fruits of the time my dad spent with me, teaching me to love math and teaching me to teach it to others.

I still have his slide rule. It's one of my most prized possessions.

Who This Book Is For

If you're interested in math, this book is for you! I've tried to write it so that it's accessible to anyone with a basic high-school background in math. The more background you have, the more depth you'll notice, but even if you've only taken high-school algebra, you should be able to follow along.

How to Read This Book

This isn't a book that you need to read cover-to-cover. Each chapter is mostly stand-alone. You can pick topics that interest you and read them in any order. Within the six parts of the book, chapters will often refer back to previous chapters in the same part for details. You'll get more out of those chapters if you read the referenced sections, but if you don't feel like it, you should still be able to follow along.

What Do You Need?

For most of the book, you need nothing but curiosity. In a few sections, there are a couple of programs. In case you want to run them, there are links and instructions in the section with the program.

Acknowledgments

It's always tricky trying to acknowledge everyone who contributed to a book like this. I'm sure that I'll wind up forgetting someone: if you deserved an acknowledgement but I left you out, I apologize in advance and thank you for your help!

Many thanks to the following people:

- My "blogfather" and friend Orac (aka David Gorski), who gave me the motivation to start my blog and helped me get the attention of readers when I was starting out

- The many readers of my blog, who've caught my mistakes and helped me become a better writer
- My fellow bloggers at Scientopia
- The people who gave time and effort doing technical reviews of drafts of this book: Paul Keyser, Jason Liszka, Jorge Ortiz, and Jon Shea
- My coworkers at Foursquare, for giving me support and feedback and for making work such a fun place to be
- The crew at The Pragmatic Bookshelf, especially David Thomas and David Kelly, who went above and beyond the call of duty to make it possible to typeset the math in this book
- And, of course, my family, for putting up with a crazed geek writer

Part I

Numbers

When you think about math, the first thing that comes to mind is probably numbers. Numbers are fascinating things. But when you think about it, it's crazy to realize just how little most of us actually understand about them.

How do you define what a number actually is? What makes a number a real number? Or a *real* number? How many numbers are there? How many different kinds of numbers are there?

I can't possibly tell you everything there is to know about numbers. That could fill twenty or thirty different books. But I can take you on a sort of a cruise, looking at some of the basic stuff about what numbers are and then looking at some of the weird and interesting numbers among them.

Part I

Numbers

[illegible]

[illegible]

[illegible]

1

Natural Numbers

What's a number?

In math, we could answer that question a few different ways. We could answer it *semantically* by looking at what numbers mean. Or we could answer that question *axiomatically* by looking at how they behave. Or we could answer the question *constructively* by looking at how we could build the numbers from some other kind of simpler object.

We'll start with semantics. What do numbers *mean*? We all think we know the answer to this, but in fact, we're wrong most of the time! People think that a number is just one thing, the thing that you use to count, and that's it. But that's not really true. Numbers can have two different meanings, depending on how they're used.

There are two kinds of numbers. When you see the number 3, you don't really know what it means. It could have two different meanings, and without knowing which one you're using, it's ambiguous. As we'll see in a moment, it could mean 3 as in "I have three apples," or it could mean 3 as in "I came in third in the race." The 3 in "three apples" is a cardinal number, and the 3 in "third place" is an ordinal number.

A *cardinal* number counts how many objects there are in a group. When I say "I want *three* apples," that three is a cardinal. An *ordinal* number counts where a particular object is in a group. When I say "I want the third apple," that three is an ordinal. In English, this distinction is easy to make because there's a specific grammatical form called the ordinal form: "three" for cardinals, "third" for ordinals, and the distinction between cardinal and ordinal is exactly the same distinction used in English grammar.

The cardinal/ordinal distinction really starts to make sense when you talk about the set theoretic basis for math. We'll look at this in more detail when we talk about set theory in 16, *Cantor's Diagonalization: Infinity Isn't Just Infinity*, on page 127. For now, this basic idea is enough: cardinals count objects; ordinals position them.

The axiomatic part is a lot more interesting. In an axiomatic definition we describe what we're looking at in terms of a collection of rules, called *axioms*. The axioms work by defining how the numbers (or whatever we're defining) behave. In math, we always prefer to work with axiomatic definitions because an axiomatic definition removes all ambiguity about what is possible and how it works. An axiomatic definition has less intuitive meaning, but it's absolutely precise, and it's structured in a way that makes it possible to use it in formal reasoning.

The Naturals, Axiomatically Speaking

We'll start by talking about the basic fundamental group of numbers: the naturals. The natural numbers (written **N**) consist of zero and the numbers greater than zero that can be written without fractions.

When you talk about numbers, you start with the natural numbers because they're the most basic fundamental sort of number. Intuitively, natural numbers are the first mathematical concepts that we understand as children. They're the whole numbers, with no fractions, starting at zero and going onward toward infinity: 0, 1, 2, 3, 4, ... (Computer scientists like me are particularly fond of natural numbers because everything computable ultimately comes from the natural numbers.)

The natural numbers are actually formally defined by a set of rules called *Peano arithmetic*. Peano arithmetic specifies a list of the axioms that define the natural numbers.

Initial Value Rule: There is one special object called 0, and 0 is a natural number.

Successor Rule: For every natural number n there is exactly one other natural number, called its *successor, $s(n)$*.

Predecessor Rule: Zero is not the successor of any natural number, and every natural number *except* 0 is the successor to some other natural number, called its *predecessor*. Say you have two numbers, a and b; if b is a's successor, then a is b's predecessor.

Uniqueness Rule: No two natural numbers have the same successor.

Equality Rules: Numbers can be compared for equality. This has three subrules: equality is reflexive, which means that every number is equal to itself; equality is symmetric, meaning that if a number a is equal to a number b, then $b = a$; and equality is transitive, which means that if $a = b$ and $b = c$, then $a = c$.

Induction Rule: For some statement P, P is true for all natural numbers if

1. P is true about 0 (that is, $P(0)$ is true).
2. If you *assume* P is true for a natural number n ($P(n)$ is true), then you can *prove* that P is true for the successor $s(n)$ of n (that $P(s(n))$ is true).

And all of that is just a fancy way of saying that the natural numbers are numbers with no fractional part starting at 0. On first encountering the Peano rules, most people find them pretty easy to understand, except for the last one. Induction is a tricky idea. I know that when I first saw an inductive proof I certainly didn't get it; it had a feeling of circularity to it that I had trouble wrapping my head around. But induction is essential: the natural numbers are an infinite set, so if we want to be able to say anything about the entire set, then we need to be able to use some kind of reasoning to extend from the finite to the infinite. That is what induction does: it lets us extend reasoning about finite objects to infinite sets.

When you get past the formalism, what the induction rule really says is, here's a pattern that you can use. If you make a definition work for your first number, then you can define it for all of the numbers after that first number by talking about what happens when you add 1 to it. With that pattern, we can write proofs about statements that are true for all natural numbers, or we can write definitions that work for all natural numbers. And using similar tricks, we can talk about all integers or all fractions or all real numbers.

Definitions are easier, so we'll do one of them before we try a proof. To give an example of how we use induction in a definition, let's look at addition. We can define addition on the natural numbers quite easily. Addition is a function "+" from a pair of natural numbers to another natural number called their *sum*. Formally, addition is defined by the following rules:

Commutativity For any pair of natural numbers n and m,

$$n + m = m + n$$

Identity For any natural numbers n,

$$n + 0 = 0 + n = n$$

Recursion For any natural numbers m and n,

$$m + s(n) = s(m + n)$$

The last rule is the inductive one, and it's built using recursion. Recursion is difficult when you're not used to it, so let's take the time to pick it apart.

What we're doing is defining addition in terms of the successor rule from Peano arithmetic. It's easier to read if you just rewrite it a tad to use +1 and –1: $m + n = 1 + (m + (n - 1))$.

What you need to remember to help make sense out of this is that it's a *definition*, not a *procedure*. So it's describing what addition means, not how to do it.

The last rule works because of the Peano induction rule. Without it, how could we define what it means to add two numbers? Induction gives us a way of saying what addition means for *any* two natural numbers.

Now for a proof. Proofs are scary to most people, but there's no need to fear them. Proofs really aren't so bad, and we'll do a really easy one.

Using Peano Induction

One simple but fun proof using the natural numbers with addition is this one: suppose I have a natural number N. What's the sum of all of the integers from 1 to N? It's N times $N + 1$ divided by 2. So how can we prove that using the induction rule?

We start with what's called a *base case*. That means we need to start with one case that we can prove all by itself, and then that case will be used as the base on which we build the induction. In the induction rule, the first clause says that the fact we want to prove needs to start by showing it for 0, so 0 is our base case. It's easy to prove this for zero: $(0*(0 + 1))/2$ is 0. So our equation is true when $N = 0$. That's it: that's our base case done.

Now comes the inductive part. Suppose for a number N that it's true. Now we want to prove it for $N + 1$. What we're going to do here is the heart of induction, and it's an amazing process. We want to show that since we know the rule is true for 0, then it must also be true for 1. Once we know it's true for 1, then it must be true for 2. If it's true for 2, then it must be true for 3. And so on. But we don't want to have to do a separate proof of each of those. So we just say, "Suppose it's true for N; then it must be true for $N + 1$." By putting the variable in this inductive structure, we're simultaneously doing "If it's true for 0, then it's true for 1; if it's true for 1, then it's true for 2, and so forth."

Here's what we want to prove:

$$(0 + 1 + 2 + 3 + \ldots + n + n + 1) = \frac{(n+1)(n+2)}{2}$$

To start, we know this:

$$(0 + 1 + 2 + 3 + \cdots + n) = \frac{n(n+1)}{2}$$

So we can substitute that in and get this:

$$\frac{n(n+1)}{2} + n + 1 = \frac{(n+1)(n+2)}{n}$$

Now we expand the multiplication on both sides:

$$\frac{n^2 + n}{2} + (n + 1) = \frac{n^2 + 3n + 2}{2}$$

Get a common denominator on the left side:

$$\frac{n^2 + n + 2n + 2}{2} = \frac{n^2 + 3n + 2}{2}$$

Finally, simplify the left side:

$$\frac{n^2 + 3n + 2}{2} = \frac{n^2 + 3n + 2}{2}$$

And that's it: we've just proven that it works for all natural numbers.

That's the axiomatic version of the naturals. They're numbers greater than or equal to zero, where each number has a successor and on which you can use that successor relationship to do induction. Pretty much everything that we do with natural numbers, and the bulk of the basic intuitive arithmetic stuff that we learn as children, can be built from nothing more than that.

After all of that, can we say what a number is? Sort of. One of the lessons of numbers in math is that numbers don't have just one meaning. There's a whole lot of different sorts of numbers: natural numbers, integers, rational numbers, real numbers, complex numbers, quaternions, surreals, hyperreals—on and on and on. But the whole universe of numbers starts with what we did right here: the natural numbers. And ultimately, the meanings of those numbers come down to either quantity or position. They're all ultimately cardinals and ordinals, or collections of cardinals and ordinals. That's what a number is: something that is *constructed from* a notion of either quantity or position.

2

Integers

Natural numbers are the first numbers that we understand. But they're nowhere close to sufficient. Given how we use numbers, you inevitably wind up needing something beyond just the natural numbers.

If you go to the store and buy something, you give them money in exchange for what you buy. You could buy something like a loaf of bread for three dollars. If you paid them with a five dollar bill, they'll give you two dollars back in change.

Just to understand that, you're doing something that doesn't really make sense in natural numbers. The money is flowing in two different directions. One, going from you to the store, subtracts from the money you have. The other, going from the store to you, adds to the money you have. Positive and negative integers allow you to distinguish between the two directions in which the money can move.

What's an Integer?

If you have the natural numbers and you want the integers, all you have to do is add an *additive inverse*. If you understand the naturals and want to understand the integers, you also only need to add one thing: *direction*. If you think of a number line, the natural numbers start from zero and go to the right, but there's nothing to the left of zero. Integers start with the naturals and add negative numbers moving to the left on the other side of zero.

The meaning of integers follows from the notion of direction. Positive integers mean the same thing as the natural numbers, both as cardinals and as ordinals. Negative numbers allow you to move in the other direction. If you think in terms of cardinals, integers allow you to talk about moving things between sets. If you have a set of size 27 and another set of size 29, then to make the sets have the same size, you can either add two objects to the first set or remove two objects from the second set. If you add to the first set, you're doing something with a positive cardinality. If you remove from the second set, you're doing something with a negative cardinality.

In ordinals, it's even easier. If you're looking at the third element in a set and you want to look at the fifth, you move forward two steps, and that motion is described by a positive ordinal integer. If you're looking at the fifth element and you want to look at the third, you move backward two steps, and that motion is a negative ordinal integer.

Let's move on to an axiomatic definition. The integers are what you get when you extend the naturals by adding an *inverse* rule. Start with the set of natural numbers, **N**. In addition to the Peano rules, we just need to add a definition of the *additive inverse*. The additive inverse of a non-zero natural number is just a negative number. To get the integers, we just add these new rules:

Additive Inverse For any natural number n other than zero, there is exactly one number $-n$ that is *not* a natural number and that is called the *additive inverse of n*, where $n + -n = 0$. We call the set of natural numbers and their additive inverses the *integers*.

Inverse Uniqueness For any two integers i and j, i is the additive inverse of j if and only if j is the additive inverse of i.

With those rules, we've got something new. The set of values that we talked about as the natural numbers can't satisfy those rules. Where do all of these new values—the negative integers—come from?

The answer is a bit disappointing. They don't come from anywhere; they just *are*. We can't create objects in math; we can only describe them. The numbers—naturals, integers, reals—exist because we define rules that describe them and because those rules describe something consistent.

All of that is just a fancy way of saying that the integers are all of the whole numbers: zero, the positives, and the negatives.

What's pretty neat is that if you define addition for the natural numbers, the inverse rule is enough to make it work for the integers as well. And since multiplication on natural numbers is just repeated addition, that means that multiplication works for the integers too.

Constructing the Integers—Naturally

We can create mathematical constructions to *represent* the integers in terms of the naturals. These constructions are called *models* of the integers. But why would we want to do that? And what is a model, precisely?

In a model of something new, like our integers, we're trying to show that there's a way of making objects that will obey the axioms that we've defined. To do that, you take the things that you already know exist and use them as building blocks. With those building blocks, you build objects that will follow the axioms of the new system. For example, to construct the integers, we're going to start with the natural numbers, which are objects that we already know about and understand. Then we're going to use them to create objects that represent the integers. If we can show that the objects in the model follow the axioms of the natural numbers, we know that our definitions of the integers are consistent.

Why should we go through all this?

There are two reasons for building constructions like models. First, a model shows us our axioms make sense. When we write a set of axioms, it's easy to screw up and accidentally write our model in an inconsistent way. A model proves that we didn't. We can write out a bunch of axioms that look reasonable but are inconsistent in a subtle way. If that's true, it means that the objects we defined don't exist, even in the abstract mathematical sense. And worse, it means that if we work as if they do exist, then every conclusion we draw will be worthless. I said earlier that the integers exist because we defined them and the definitions are consistent. If we don't show that it's possible to create a model, then we can't be sure that the definitions really are consistent.

The other reason is less abstract: a model makes it easier for us to understand and to describe how the system we are building is supposed to work.

One last caveat before we get to the model. It's really important to understand that what we're doing is creating *a* model of the integers, not *the* model of the integers! What we're doing here is describing one possible way of representing the integers. The integers are not the representation that I'm going to show you. There are many possible representations; so long as a representation fits the axioms, it can be used. The distinction between the model and the thing that it models is subtle, but it's very important. The integers are the objects described by the axioms, not the model that we're building. The model is just a representation.

The simplest way of representing the integers is to use ordered pairs of natural numbers: (a, b). The integer represented by a pair (a, b) is the value of $(a - b)$. Obviously then, $(2, 3)$, $(3, 4)$, $(18, 19)$ and $(23413, 23414)$ are all representations of the same number. In mathematical terms, we say that the integers consist of equivalence classes of those pairs.

But what's an equivalence class?

When we're doing things like building a model of the integers, very often the way we define them doesn't create exactly one object for each integer. We define the model so that for each object in the thing we're modeling, there's a collection of values that are equivalent within the model. That group of equivalent values is called an equivalence class.

In our model of the integers, we're creating pairs of natural numbers to model each specific integer. Two of those pairs (a, b) and (b, c) are equivalent when the first and second element of each pair are separated by the same distance and direction on a number line, such as $(4, 7)$ and $(6, 9)$. On a number line, to go from 4 to 7, you have to take three steps to the right. To go from 6 to 9, you still have to go three steps to the right, so they're in the same equivalence class. But if you look at $(4, 7)$ and $(9, 6)$, you'd go three steps to the right to get from 4 to 7, but three steps to the left to get from 9 to 6. So they are *not* in the same equivalence class.

The representation here gives us an easy way to understand what the various arithmetic operations mean when applied to the integers in terms of what those operations mean on the natural numbers. We understand what addition means on the natural numbers, and so we can use that to define addition on the integers.

If you have two objects from our model of the integers, they're defined as pairs of natural numbers: $M = (m_1, m_2)$ and $N = (n_1, n_2)$. Addition and subtraction on them are defined by the following rule:

- $M + N = (m_1 + n_1, m_2 + n_2)$
- $M - N = (m_1 + n_2, m_2 + n_1)$
- The additive inverse of a number $N = (n_1, n_2)$, written $-N$, is just the reverse pair: $-N = (n_2, n_1)$.

The definition of subtraction turns out to be pretty neat. $3 - 5$ would be $(3, 0) - (5, 0)$, which is equal to $(3, 0) + (0, 5) = (3, 5)$, which is an element of the equivalence class of -2. And the definition of additive inverse is just a natural result of how we define subtraction: $-N = 0 - N$.

That's all we need to do to get from the natural numbers to the integers: just add additive inverses. We could do subtraction with just the naturals, which would almost require additive inverses in some sense, but it would be messy.

The problem would be that in just the natural numbers, you *can't* define subtraction as an operation on any two values. After all, if you subtract 5 from 3, the result can't be defined with just the natural numbers. But with the integers, subtraction becomes a really general operation: for any two integers M and N, $M - N$ is another integer. In formal terms, we say that subtraction is a *total function* on the integers and that the integers are *closed* over subtraction.

But that leads us toward the next problem. When we look at addition in the integers, it has a natural inverse operation in subtraction, which can be defined in terms of the additive inverse of the integers. When we move on to the next common operation, multiplication, we can define multiplication on the naturals and the integers, but we can't define its inverse operation, division, because there is no way that we can define a multiplicative inverse that will work on the integers. To describe division as a well-defined operation, we need to have another kind of number—the rational numbers—which we'll talk about next.

3

Real Numbers

Now we know about the naturals and the integers. That's a good start. But there are more kinds of numbers than that: there are fractions and irrationals and…well, we'll come to that later. But the next step in understanding numbers is looking at numbers that have non-integer parts, numbers that fit into the gaps between integers, like 1/2, –2/3, and π.

For now, we'll look at the next kind of numbers—numbers with a non-integer part, otherwise known as the *real* numbers.

Before I go into detail, I need to say up front that I hate the term "real numbers." It implies that other kinds of numbers are not real, which is silly, annoying, frustrating, and not true. In fact, the term was originally coined as a counterpart to *imaginary* numbers, which we'll talk about in 8, *i: The Imaginary Number,* on page 47. Those were named "imaginary" as a way of mocking the concept. But the term *real numbers* has become so well established that we're pretty much stuck with it.

There are a couple of ways to describe the real numbers. I'm going to take you through three of them: first, an informal intuitive description; then, an *axiomatic* definition; and finally, a *constructive* definition.

The Reals, Informally

An informal and intuitive way to depict the real numbers is the number line that we learned in elementary school. Imagine a line that goes on forever in both directions. You can pick a spot on it and label it 0. To the right of 0, you mark off a second spot and label it 1. The distance between 0 and 1 is the distance between any two adjacent integers. So go the same distance to the right, make another mark, and label it 2. Keep doing that as far as you want. Then start going to the left from 0. The first mark is –1, the second –2, etc. That's a basic number line. I've drawn a version of that in the following figure. Anyplace on that line that you look is a *real* real number. Halfway between 0 and 1 is the real number 1/2. Halfway between 0 and 1/2 is 1/4. You can keep dividing it forever: between any two real numbers you can always find another real number.

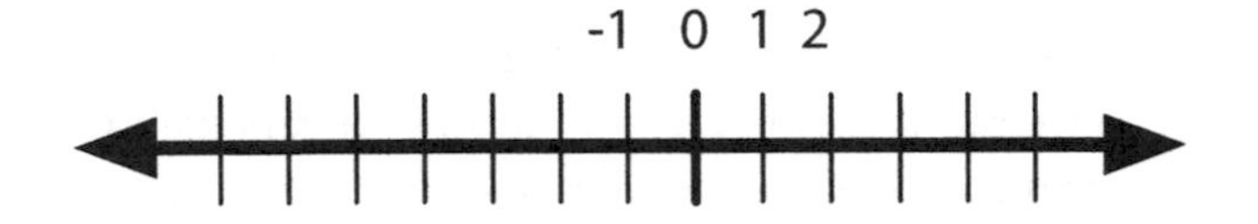

Figure 1—The number line. The real numbers can be represented by the points to the left and right of 0 on an infinitely long line.

Using the number line, most of the important properties of the real numbers can be described in very nice, intuitive ways. The ideas of addition, subtraction, ordering, and continuity are all very clear. Multiplication may seem tricky, but it can also be explained in terms of the number line (you can look at my blog for posts about slide rules to get an idea of how[1]).

What the number line gives us to start isn't quite the real numbers. It's the *rational* numbers. The rationals are the set of numbers that can be expressed as simple fractions: they're the *ratio* of one integer to another. ½, ¼, ⅗, 124342/58964958. When we look at the number line, we usually think about it in terms of the rationals. Consider for a moment the way that I described the number line a couple of paragraphs ago: "You can keep dividing it forever: between any two real numbers, you can always find another real number." That division process will always give us a rational number.

1. http://scientopia.org/blogs/goodmath/2006/09/manual-calculation-using-a-slide-rule-part-1

Divide any fraction into any number of equal parts, and the result is still a fraction. No matter how many times we divide using rationals and integers, we can never get anything that isn't another rational number.

But even with the rational numbers, there are still *gaps* that can't be filled. (We know about some of the numbers that fit into those gaps—they're irrational numbers like the familiar π and e. We'll see more about the irrational numbers in 4, *Irrational and Transcendental Numbers*, on page 23, and about e specifically in 6, *e: The Unnatural Natural Number*, on page 37.) Looking at the rationals, it's hard to see how there can be gaps. No matter what you do, no matter how small the distance between two rational numbers is, you can always fit an infinite number of rational numbers between them. How can there be gaps? The answer is that we can easily define sequences of values that have a limit but whose limit cannot possibly be a rational number.

Take any finite collection of rational numbers and add them up. Their sum will be a rational number. But you can define infinite collections of rational numbers, and when you add them up, the result *isn't* a rational number! Here's an example:

$$\pi = \frac{4}{1} - \frac{4}{3} + \frac{4}{5} - \frac{4}{7} + \frac{4}{9} - \frac{4}{11} + \cdots$$

Every term of this sequence is obviously a rational number. If you work out the result of the first two terms, then the first three, then the first four, and so on, you get 4.0, 2.666..., 3.4666..., 2.8952..., 3.3396... and after 100,000 terms, it's about 3.14158. If you keep going, it's clearly converging on a specific value. But no finite sequence of rationals will ever be exactly the limit of that sequence. There's something that is the limit of this sequence, and it's clearly a number; and no matter what we do, it's never exactly equal to a rational number. It's always somewhere between any two rational numbers that we can choose.

The real numbers are the integers, the rational numbers, and those strange numbers that fit into the gaps between the rational numbers.

The Reals, Axiomatically

The axiomatic definition is, in many ways, quite similar to the definition the number line provides, but it does the job in a very formal way. An *axiomatic* definition doesn't tell you how to get the real numbers; it just describes them with rules that draw on simple set theory and logic.

When we're building something like the real numbers, which are defined by a set of related components, mathematicians like to be able to say that what we're defining is a single *object*. So we define it as a *tuple*. There's no deep meaning to the construction of a tuple; it's just a way of gathering components into a single object.

The reals are defined by a tuple: $(\mathbf{R}, +, 0, \times, 1, \leq)$, where **R** is an infinite set, "+" and "×" are binary operators on members of **R**, "0" and "1" are special distinguished elements of **R**, and "≤" is a binary relation over members of **R**.

The elements of the tuple must satisfy a set of axioms, called the *field axioms*. The real numbers are the canonical example of a mathematical structure called a *field*. A field is a fundamental structure used all over the place in mathematics; it's basically the structure that you need to make algebra work. We define a field axiomatically by a set of *field axioms*. The field axioms get a bit hairy, so instead of looking at them all at once, we'll go through them one by one in the following sections.

Field Axioms Part 1: Addition and Multiplication

Let's start with the most basic axioms. The real numbers (and all fields) have two main operations: addition and multiplication. Those two operations need to work together in specific ways:

- $(\mathbf{R}, +, \times)$ are a field. Here's what this means:
  - "+" and "×" are closed, total, and onto in **R**. *Closed* means that for any pair of real numbers r and s, if you add them together, $r + s$ and $r \times s$ will be real numbers. *Total* means that for any possible pair of real numbers r and s, you can add $r + s$ or multiply $r \times s$. (It may sound silly to say that, but remember:

we're going to get to division soon, and for division, that's not true. You can't divide by zero.) And finally, *onto* means that if you have any real number x, you can find pairs of real numbers like r and s or t and u, where $r + s = x$ and $t \times u = x$.

– "+" and "×" are commutative: $a + b = b + a$, $a \times b = b \times a$.

– "×" is distributive with respect to each "+." That means that $(3 + 4) \times 5 = 3 \times 5 + 4 \times 5$.

– 0 is the only identity value for "+." For all a, $a + 0 = a$.

– For every member x of the set **R**, there is *exactly one* value $-x$, called the *additive inverse* of x, so that $x + -x = 0$, and for all $x \neq 0$, $x \neq -x$.

– 1 is the only identity value for "×"; for all a, $a \times 1 = a$.

– For every real number x except 0, there is exactly one value x^{-1}, called the *multiplicative inverse* of x, such that $x \times x^{-1} = 1$; and unless x is 1, x and x^{-1} are not equal.

If you turn all of that into English, it's really not hard. It just says that addition and multiplication work the way that we learned in school. The difference is that in school, we were taught that this is just how numbers work; now we're stating the axioms explicitly as requirements. The real numbers are the real numbers *because* they work this way.

Field Axioms Part 2: Ordering

The next axioms are about the fact that the real numbers are ordered. Basically, it's a formal way of saying that if you have two real numbers, one is less than the other unless they're equal.

- *(**R**, ≤)* is a total order:

  1. For all real numbers a and b, either $a \leq b$ or $b \leq a$ (or both, if $a = b$).

  2. "≤" is transitive: if $a \leq b$ and $b \leq c$ then $a \leq c$.

  3. "≤" is antisymmetric: if $a \leq b$, and $a \neq b$, then it's not true that $b \leq a$.

- "≤" is compatible with "+" and "×":
  1. If $x \leq y$ then $(x + 1) \leq (y + 1)$.
  2. If $x \leq y$, then for all z where $0 \leq z$, $(x \times z) \leq (y \times z)$.
  3. If $x \leq y$, then for all $z \leq 0$, $(x \times z) \leq (y \times z)$.

Field Axioms Part 3: Continuity

Now, we get to the hard one. The tricky bit about the real numbers is the fact that they're continuous—meaning that given any two real numbers, there's an infinite number of real numbers between them. And in that infinitely large collection of reals, the total order still holds. To say that, we have to talk in terms of *upper bounds*:

- For every subset S in **R** where S is not the empty set, if S has an upper bound, then it has a *least upper bound, l,* such that for any real number x that is an upper bound for S, $l \leq x$.

That really says this: if you take a bunch of real numbers, no matter how close together or how far apart they are, there is one number that is the smallest number that's larger than all of the members of that set.

That's an extremely concise version of the axiomatic definition of reals. It describes what properties the real numbers must have, in terms of statements that could be written out in a formal, logical form. A set of values that match that description is called a model for the definition; you can show that there are models that match the definition and that all of the models that match the definition are equivalent.

The Reals, Constructively

Finally, the *constructive* definition—a constructive definition is a procedure for creating the set of real numbers. We can think of the real numbers as the union of several different sets.

First, we'll take the integers. All of the integers are real numbers, with exactly the properties they had as integers.

Then we'll add fractions, which are formally called *rational* numbers. A rational number is defined by a pair of non-zero integer numbers called a *ratio*. A ratio n/d represents a real number that when multiplied by d gives the value n. The set of numbers constructed this way ends up with lots of equivalent values, such as 1/2, 2/4, 3/6, and so on. As we did with the integers, we'll define the rationals as a set of *equivalence classes* over the ratios.

Before we can define the equivalence classes of the rational numbers, we need to define a couple of other things that we'll need:

1. If (a/b) and (c/d) are rational numbers, then $(a/b) \times (c/d) = (a \times c)/(b \times d)$.

2. For every rational number except 0, there's another rational called its *multiplicative inverse*. If a/b is a rational number, then its multiplicative inverse, written $(a/b)^{-1}$, is (b/a). For any two rational numbers x and y, if $y = x^{-1}$ (if y is the multiplicative inverse of x), then $x \times y = 1$.

We can use the definition of the multiplicative inverse to define ratio equivalence. Two ratios a/b and c/d are equivalent if $(a/b) \times (c/d)^{-1} = 1$; that is, if multiplying the first ratio by the multiplicative inverse of the second ratio is 1. The equivalence classes of ratios are the rational numbers, and every rational number is also a real number.

That gives us the complete set of the rational numbers. For convenience, we'll use **Q** to represent the set of rational numbers. Now we're kind of stuck. We know that there are irrational numbers. We can define them axiomatically, and they fit the axiomatic definition of reals. But we need to be able to construct them. How?

Mathematicians have a bunch of tricks at their disposal that they can use to construct the real numbers. The one I'm going to use is based on something called a *Dedekind cut*. A Dedekind cut is a mathematical object that can represent a real number r as a pair (A, B) of sets: A is the set of rational numbers smaller than r; B is the set of rational numbers larger than r. Because of the nature of the rationals, these two sets have really peculiar properties. The set A is a set containing values *smaller than* some number r; but there is no *largest value* of A. B is similar: there is no smallest value in B. r is the number in the gap between the two sets in the cut.

How does that get us the irrational numbers? Here's a simple example: let's define the square root of 2 using a Dedekind cut:

$$A = \{ r : r \times r < 2 \text{ or } r < 0 \}$$
$$B = \{ r : r \times r > 2 \text{ and } r > 0 \}$$

Using Dedekind cuts makes it easy to define the real numbers constructively. We can say that the set of real numbers is the set of numbers that can be defined using Dedekind cuts of the rationals.

We know that addition, multiplication, and comparisons work nicely on the rationals—they form a field and they are totally ordered. Just to give you a sense of how we can show that the cuts also fit into that, we can show the definitions of addition, equality, and ordering in terms of cuts.

- *Addition*: This shows the sum $X + Y$ of two cuts, $X = (X_L, X_R)$ and $Y = (Y_L, Y_R) = (Z_L, Z_R)$, where $Z_L = \{ x + y : x \text{ in } X_L \text{ and } y \text{ in } Y_L\}$ and $Z_R = \{ x + y : x \text{ in } X_R \text{ and } y \text{ in } Y_R\}$.
- *Equality*: Two cuts $X = (X_L, X_R)$ and $Y = (Y_L, Y_R)$ are equal if and only if X_L is a subset of Y_L and X_R is a subset of Y_R.
- *Ordering*: If you have two cuts, $X = (X_L, X_R)$ and $Y = (Y_L, Y_R)$, X is less than or equal to Y if and only if X_L is a subset of Y_L and Y_R is a subset of X_R.

Now that we've defined the reals and shown how we can construct good models of them, we've gotten to the point where we know what the most familiar numbers are and how they work mathematically. It might seem like now we really understand numbers. But the fact is, we don't really. Numbers are still full of surprises. To give you a very quick taste of what's coming, it turns out that we can't *write* most numbers. Most numbers go on forever and ever, and we can't write them down. In fact, most numbers, we can't even give a name to. We can't write a computer program that will find them. They're real and they exist; but we can't point at them, name them, or describe them individually! In the next section, we'll look at a class of numbers called the *irrational numbers* that are the root of this uncomfortable fact.

4

Irrational and Transcendental Numbers

In the history of math, there've been a lot of disappointments for mathematicians. They always start off with the idea that math is a beautiful, elegant, perfect thing. They pursue it, and they eventually discover that it's not.

This leads us to a collection of strange numbers that we need to deal with: the irrational and transcendental numbers. Both were huge disappointments to the mathematicians who discovered them.

What Are Irrational Numbers?

Let's start with the irrational numbers. These are numbers that aren't integers and also aren't a ratio of any two integers. You can't write them as a normal fraction. If you write them as a continued fraction (which we'll describe in 11, *Continued Fractions*, on page 69), then they go on forever. If you write them in decimal form, they go on forever without repeating. They're called irrational because they can't be written as ratios. Many people have claimed that they're irrational because they don't make sense, but that's just a rationalization after the fact.

They do make sense, but they are uncomfortable and ugly to many mathematicians. The existence of irrational numbers means that there are numbers that you cannot write down, and that's an unpleasant fact. You can't ever be precise when you use them: you're always using approximations because you can't write them down exactly. Any time you do a calculation using a representation of an irrational number, you're doing an approximate calculation, and you can only get an approximate answer. Unless you manipulate them symbolically, no calculation that involves them can ever be solved exactly. If you're looking for perfection—for a world in which numbers are precise and perfect—this isn't it.

The transcendental numbers are even worse. *Transcendental numbers* are irrational; but not only can transcendental numbers not be written as a ratio of integers, not only do their decimal forms go on forever without repeating, transcendental numbers are numbers that can't be described by algebraic operations. There are irrational numbers like the square root of 2, which you can easily define in terms of an algebraic equation: it's the value of x in the equation $y = x^2 - 2$ where $y = 0$. You can't write the square root of 2 as a decimal or a fraction, but you can write it with that simple equation. When you're looking at a transcendental number, you can't even do that. There's no finite sequence of multiplications, divisions, additions, subtractions, exponents, and roots that will give you the value of a transcendental number. The square root of 2 is not transcendental, because you can describe it algebraically; but e is.

The *Argh!* Moments of Irrational Numbers

According to legend, the first disappointment involving the irrational numbers happened in Greece around 500 BC. A rather brilliant man by the name of Hippasus, who was part of the school of Pythagoras, was studying roots. He worked out a geometric proof of the fact that the square root of 2 could not be written as a ratio of integers. He showed it to his teacher, Pythagoras. Pythagoras, like so many other mathematicians, was convinced that numbers were clean and perfect and he could not accept the idea of irrational numbers. After analyzing Hippasus's proof and being unable to find any error in it, he became so enraged that he *drowned* poor Hippasus.

A few hundred years later, Eudoxus worked out the basic theory of irrationals, and it was published as a part of Euclid's mathematical texts.

From that point, the study of irrationals pretty much disappeared for nearly two thousand years. It wasn't until the seventeenth century that people really started looking at them again. And once again, it led to disappointment, but at least no one got killed this time.

With the acceptance of irrational numbers, the idea of numbers as something that allowed us to capture the world precisely fell apart. Even something like calculating the circumference of a circle couldn't be done precisely. But mathematicians didn't give up on perfection. They came up with a new idea for what the perfection of numbers in mathematics meant, this time based on algebra. This time they theorized that while you might not be able to write down all numbers as ratios, all numbers must be describable using algebra. Their idea was that for any number, whether integer, rational, or irrational, there was a finite polynomial equation using rational coefficients that had the number as a solution. If they were correct, then any irrational number could be computed by a finite sequence of addition, subtraction, multiplication, division, exponents, and roots.

But it was not to be. The German philosopher, mathematician, and man about town Gottfried Wilhelm Leibniz (1646–1716) was studying algebra and numbers, and he's the one who made the unfortunate discovery that lots of irrational numbers are algebraic but lots of them aren't. He discovered it indirectly by way of the sine function. Sine is one of the basic operations of trigonometry, the ratio of two sides of a right triangle. The sine function is one of the fundamentals of analytic geometry that has real-world implications and is not just a random weird function that someone made up. But Leibniz discovered that you couldn't compute the sine of an angle using algebra. There's no algebraic function that can compute it. Leibniz called sine a transcendental function, since it went beyond algebra. This wasn't quite a transcendental *number*, but it really introduced the idea that there were things in math that couldn't be done with algebra.

Building on the work of Leibniz, the French mathematician Joseph Liouville (1809–1882) worked out that you could easily construct numbers that couldn't be computed using algebra. For example, the constant named after Liouville consists of a string of 0s and 1s where for digit x, 10^{-x} is a 1 if and only if there is some integer n such that $n! = x$.

Once again, mathematicians tried to salvage the beauty of numbers. They came up with a new theory: that transcendental numbers existed, but they needed to be *constructed.* They theorized that while there were numbers that couldn't be computed algebraically, they were all contrived things, things that humans designed specifically to be pathological. They weren't *natural*.

Even that didn't work. Not too much later, it was discovered that e was transcendental. And as we'll see in 6, *e: The Unnatural Natural Number,* on page 37, e is a natural, unavoidable constant. It is absolutely not a contrived creation. Once e was shown to be transcendental, other numbers followed. In one amazing proof, π was shown to be transcendental *using* e. One of the properties that they discovered after recognizing that e was transcendental was that any transcendental number raised to a non-transcendental power was transcendental. Since the value of $e^{i\pi}$ is not transcendental (it's –1), then π must be transcendental.

An even worse disappointment in this area came soon. One of the finest mathematicians of the age, Georg Cantor (1845–1918) was studying the irrationals and came up with the infamous "Cantor's diagonalization," which we'll look at in 16, *Cantor's Diagonalization: Infinity Isn't Just Infinity,* on page 127, which shows that there are more transcendental numbers than there are algebraic ones. Not only are there numbers that aren't beautiful and that can't be used in precise computations, but *most* numbers aren't beautiful and can't be used in precise computations.

What Does It Mean, and Why Does It Matter?

Irrational and transcendental numbers are everywhere. Most numbers aren't rational. Most numbers aren't even algebraic. That's a very strange notion: we can't write most numbers down.

Even stranger, even though we know, per Cantor, that most numbers are transcendental, it's incredibly difficult to prove that any particular number is transcendental. Most of them are, but we can't even figure out which ones!

What does that mean? That our math-fu isn't nearly as strong as we like to believe. Most numbers are beyond us. Here are some interesting numbers that we know are either irrational or transcendental:

- e: transcendental
- π: transcendental
- The square root of 2: irrational, but algebraic
- The square root of x, for all x that are not perfect squares: irrational
- $2^{\text{square root of 2}}$: irrational
- Ω, Chaitin's constant: transcendental

What's interesting is that we really don't know very much about how transcendentals interact; and given the difficulty of proving that something is transcendental, even for the most well-known transcendentals, we don't know much of what happens when you put them together. $\pi+e$; $\pi\times e$; π^e, e^e are all numbers we *don't know* are transcendental. In fact, for $\pi + e$, we don't even know if it's irrational!

That's the thing about these numbers. We have such a weak grasp of them that even things that seem like they should be easy and fundamental, we just do not know how to do. And as we keep studying numbers, it doesn't get any better. For the people who want numbers to make sense, the disappointments keep coming. Not too long ago, an interesting fellow (and former coworker of mine) named Gregory Chaitin (1947–), showed that the irrational numbers are even worse than we thought. Not only are most numbers not rational, not only are most numbers not algebraic, most numbers cannot even be *described in any way*. It's not a big surprise that they can't be written down, because we already know that we can't really write down any irrational number—the best we can do is write a good approximation. In fact, for most numbers, we can't write a description, an equation, or a computer program to generate them. We can't identify them precisely enough to name them. We know they exist, but we're absolutely helpless to describe or identify them in any way at all. It's an amazing idea. If you're interested in it, I highly recommend reading Greg's book, *The Limits of Mathematics [Cha02].*

Part II

Funny Numbers

When we think about numbers, even if we're thinking abstractly, we don't usually think of anything like an axiomatic definition such as Peano arithmetic. We think about specific numbers and their symbols.

From a mathematical viewpoint, some numbers tell us important things. For example, the number zero, which for eons wasn't even considered a number, practically changed the entire world once people understood it!

In this part of the book, we're going to look at some of the special numbers used by mathematicians and scientists because they tell us interesting things about the world and how we, as human beings, understand it. Because their properties are sometimes startling, I call them *funny* numbers.

5

Zero

When we look at strange numbers, the starting place has to be zero. Zero may not seem strange to you because you're used to it. But the idea of zero really is strange. Think about what we said numbers mean. If you think in terms of cardinals and ordinals, in terms of counting and position, what does zero mean?

As an ordinal, what does it mean to be the zeroth object in a collection? And what about zero as a cardinal? I can have one something and count it. I can have 10 somethings and count them. But what does it mean to have zero of something? It means that I don't have any of it. So how can I count it?

And yet, without the concept of zero and the numeral 0, most of what we call math would just fall apart.

The History of Zero

In our pursuit of the meaning of zero, let's start with a bit of history. Yes, there's an actual history to zero!

If we were to go back in time and look at when people started working with numbers, we'd find that they had no concept of zero. Numbers really started out as very practical tools, primarily for measuring quantity. They were used to answer questions like "How much grain do we have stored away?" and "If we eat this much now, will we have enough to plant crops next season?" When you think about using numbers in a context like that, a *measurement* of zero doesn't really mean much. A measurement can only make sense if there's something to measure.

Even when math is applied to measurements in modern math, leading zeros in a number—even if they're measured—don't count as significant digits in the measurement. (In scientific measurement, *significant digits* are a way of describing how precise a measurement is and how many digits you can use in computations. If your measurement had two significant digits, then you can't have more than two meaningful digits in the result of any computation based on that measurement.) If I'm measuring some rocks and one weighs 99 grams, then that measurement has only two significant digits. If I use the same scale to weigh a very slightly larger rock and it weighs 101 grams, then my measurement of the second rock has three significant digits. The leading zeros don't count.

We can understand early attitudes about zero by looking back to Aristotle (384–322 BC). Aristotle was an ancient Greek philosopher whose writings are still studied today as the foundations of the European intellectual tradition. Aristotle's thoughts on zero are a perfect example of the reasoning behind why zero wasn't part of most early number systems. He saw zero as a counterpart to infinity. Aristotle believed that both zero and infinity were pure ideas related to the concept of numbers and counting, but that they were not actually numbers themselves.

Aristotle also reasoned that, like infinity, you can't ever get to zero. If numbers are quantities, he thought, then obviously, if you start with one of something and cut it in half, you'll be left with half as much. If you cut it in half again, you'll have one quarter. Aristotle and his contemporaries thought that you could continue that halving process forever: 1/4, 1/8, 1/16, and so on. The amount of stuff you'll have left will get smaller and smaller, closer and closer to zero, but you'll never actually get there.

Aristotle's view of zero does make sense. After all, you can't really have zero of anything, because zero of something is nothing. When you have zero, you don't have a real quantity of stuff. Zero is the absence of stuff.

The first real use of zero wasn't really as a number, but as a digit symbol in numeric notation. The Babylonians had a base-60 number system. They had symbols for numbers from one to 60. For numbers larger than 60, they used a positional system like our decimal numbers. In that positional system, for digit-places with no number, they left a space; that space was their zero. This introduced the idea of a zero as a recordable quantity in some contexts. Later they adopted a placeholder that looked like a pair of slashes (//). It was never used by itself but only as a marking inside multidigit numbers. If the last digit of a number was zero, they didn't write it, because the zero marker was just a placeholder between two non-zero digits to show that there was something in between them. So, for example, the numbers 2 and 120 (in Babylonian base-60, that's 2 × 1 versus 2 × 60) looked exactly the same; you needed to look at the context to see which it was, because they wouldn't write a trailing zero. They had the concept of a notational zero, but only as a separator.

The first real zero was introduced to the world by an Indian mathematician named Brahmagupta (598–668) in the seventh century. Brahmagupta was quite an accomplished mathematician: he didn't just invent zero, but arguably he also invented the idea of negative numbers and algebra! He was the first to use zero as a real number and the first to work out a set of algebraic rules about how zero and positive and negative numbers worked. The formulation he worked out is very interesting; he allowed zero as both a numerator or a denominator in a fraction.

From Brahmagupta, zero spread west (to the Arabs) and east (to the Chinese and Vietnamese). Europeans were just about the last to get it; they were so attached to their wonderful roman numerals that it took quite a while to penetrate: zero didn't make the grade in Europe until about the thirteenth century, when Fibonacci (he of the sequence) translated the works of a Persian mathematician named al-Khwarizmi (from whose name sprung the word *algorithm* for a mathematical procedure). Europeans called the new number system *Arabic* and credited it to the Arabs. As we've seen, the Arabs didn't create Arabic numbers, but it was Arabic scholars, including the famous Persian poet Omar Khayyam (1048–1131), who adopted Brahmagupta's notions *and extended them* to include complex numbers, and it was their writings that introduced these ideas to Europe.

An Annoyingly Difficult Number

Even now, when we recognize zero as a number, it's an annoyingly difficult one. It's neither positive nor negative; it's neither prime nor compound. If you include it in the set of real numbers, then the fundamental mathematical structures like groups that we use to define how numbers apply to things in the world won't work. It's not a unit. Units don't work with it—for any other number, 2 inches and 2 yards mean different things—but that's not true with zero. In algebra, zero breaks a fundamental property called *closure*: without 0, any arithmetic operation on numbers produces a result that is a number. With zero, that's no longer true, because you can't divide by zero. Division is closure for every possible number *except* zero. It's a real obnoxious bugger in a lot of ways. One thing Aristotle was right about: zero is a kind of counterpart to infinity: a concept, not a quantity. But infinity we can generally ignore in our daily lives. Zero we're stuck with.

Zero is a real, inescapable part of our entire concept of numbers. But it's an oddball, the dividing line that breaks a lot of rules. For example, addition and subtraction aren't closed without zero. Integers with addition form a mathematical structure called a *group*—which we'll talk more about in 20, *Group Theory: Finding Symmetries with Sets*, on page 167—that defines what it means for something to be symmetric like a mirror reflection. But if you take away 0, it's no longer a group, and you can no longer define mirror symmetry. Many other concepts in math crumble if we take away zero.

Our notation for numbers is also totally dependent on zero; and it's hugely important for making a polynomial number system work. To get an idea of how valuable it is, just think about multiplication. Without 0, multiplication becomes much, much harder. Just compare long multiplication the way we do it with the way the Romans did multiplication, which I explain in Section 9.3, *Arithmetic Is Easy (But an Abacus Is Easier)*, on page 58.

Because of the strangeness of zero, people make a lot of mistakes involving it.

For example, here's one of my big pet peeves: based on that idea that zero and infinity are relatives, a lot of people believe that one divided by zero is infinity. It isn't. *1/0* doesn't equal anything; the way that we define what division means, it's *undefined*—the written expression *1/0* is a meaningless, invalid expression. You can't divide by 0.

An intuition supporting the fact that you can't divide by zero comes from the Aristotelean notion that zero is a concept, not a quantity. Division is a concept based on quantity, so asking "What is *X* divided by *Y*?" is asking "What quantity of stuff is the right size so that if I take *Y* of it, I'll get *X*?"

If we try to answer that question, we see the problem: what quantity of apples can I take zero of to get one apple? The question makes no sense, and it shouldn't make sense, because dividing by zero makes no sense: *it's meaningless*.

Zero is also at the root of a lot of silly mathematical puzzles and tricks. For example, there's a cute little algebraic pun that can show that 1 = 2, which is based on hiding a division by zero.

Trick: Use Hidden Division by Zero to Show That 1=2.

1. Start with $x = y$.
2. Multiply both sides by x: $x^2 = xy$.
3. Subtract y^2 from both sides: $x^2 - y^2 = xy - y^2$.
4. Factor: $(x + y)(x - y) = y(x - y)$.
5. Divide both sides by the common factor $(x - y)$, giving $x + y = y$.
6. Since $x = y$, we can substitute y for x: $y + y = y$.
7. Simplify: $2y = y$.
8. Divide both sides by y: $2 = 1$.

The problem, of course, is step 5. Because $x - y = 0$, step 5 is equivalent to dividing by zero. Since that's a meaningless thing to do, everything based on getting a meaningful result from that step is wrong—and so we get to "prove" false facts.

Anyway, if you're interested in reading more, the best source of information that I've found is an online article called "The Zero Saga."[1] It covers a bit of history and random chit-chat like this section, but it also provides a detailed presentation of everything you could ever want to know, from the linguistics of the words "zero" and "nothing" to cultural impacts of the concept, to a detailed mathematical explanation of how zero fits into algebras and topologies.

1. http://home.ubalt.edu/ntsbarsh/zero/ZERO.HTM

6

e: The Unnatural Natural Number

Our next funny number is called e, also known as Euler's constant, also known as the base of the natural logarithm. e is a very odd number, but it's also very fundamental. It shows up constantly and in all sorts of strange places where you wouldn't expect it.

The Number That's Everywhere

What is e?

e is a *transcendental* irrational number. It is roughly 2.718281828459045. It's also the base of the natural logarithm. That means that by definition, if $ln(x) = y$, then $e^y = x$.

Given my highly warped sense of humor and my love of bad puns (especially bad geek puns), I like to call e the unnatural natural number. It's natural in the sense that it's the base of the natural logarithm; but it's not a natural number according to the usual definition of natural numbers. (Hey, I warned you that it was going to be a bad geek pun!)

But that's not a sufficient answer. We call it the *natural* logarithm. Why is a bizarre irrational number that's just a bit smaller than 2 ¾ considered *natural*?

The answer becomes clear once you understand where it comes from. Take the curve $y = 1/x$. For any value n, the area under the curve from 1 to n is the natural log of n. e is the point on the x-axis where the area under the curve from 1 to n is equal to 1, as shown in the following figure. That means that the natural logarithm of a number is directly related to that number through its reciprocal.

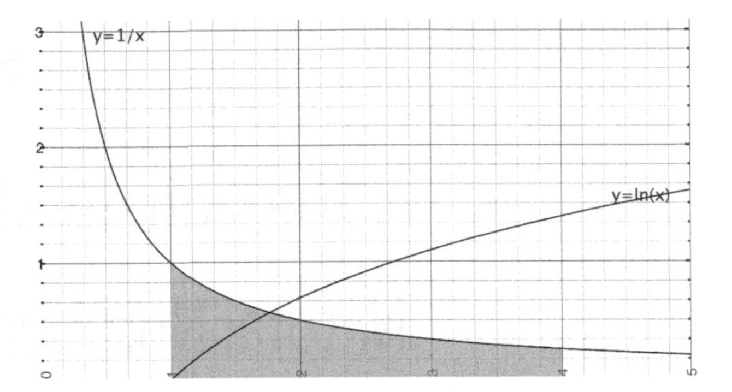

Figure 2—Finding e graphically: *The natural logarithm of a number* n *is the area under the curve from 1 to* n.

e is also what you get if you add up the reciprocal of the factorials of every natural number:

$$e = (\frac{1}{0!} + \frac{1}{1} + \frac{1}{2} + \frac{1}{3!} + \frac{1}{4!} + \cdots)$$

It's also what you get if you take this limit:

$$e = \lim_{n \to \infty}(1 + \frac{1}{n})^n$$

It's also what you get if you work out this very strange-looking series:

$$e = 2 + \cfrac{1}{1 + \cfrac{1}{2 + \cfrac{2}{3 + \cfrac{3}{(4 + \cdots)}}}}$$

It's also the base of a very strange equation in calculus:

$$\frac{de^x}{dx} = e^x$$

That last one means that e^x is its own derivative, which is more than a little strange. There's no other exponential equation that is precisely its own derivative.

And finally, it's the number that makes the most amazing equation in all of mathematics work:

$$e^{i\pi} + 1 = 0$$

That's an astonishing equation. It's taking a collection of the most fundamental and, frankly, mysterious numbers in all of mathematics and connecting them. What does it mean? We'll talk about that in 8, *i: The Imaginary Number*, on page 47.

Why does *e* come up so often? It's part of the fundamental structure of numbers. It is a deeply natural number that is a part of many of the most basic mathematical structures, like the shape of a circle. There are dozens of different ways of defining it because it's so deeply embedded in the structure of *everything*. Wikipedia even points out that if you put $1 into a bank account paying 100% interest compounded continuously, at the end of the year, you'll have exactly *e* dollars.[1] (That's not too surprising; it's just another way of stating the integral definition of *e*, but it's got a nice intuitiveness to it.)

History

As major mathematical constants go, *e* has less history to it than most. It's a comparatively recent discovery.

The first reference to it was by the English mathematician William Oughtred (1575–1660) during the seventeenth century. Oughtred is the guy who invented the slide rule, which works on logarithmic principles. The moment you start looking at logarithms, you'll start seeing *e*. Oughtred didn't actually name it, or even really work out its value, but he did write the first table of the values of the natural logarithm.

Not too much later, it showed up in the work of Gottfried Leibniz (1646–1716). Leibniz's discovery of the number wasn't too surprising, given that Leibniz was in the process of working out the basics of differential and integral calculus, and *e* shows up all the time in calculus. But Leibniz didn't call it *e*; he called it *b*.

The first person to really try to calculate a value for *e* was Daniel Bernoulli (1700–1782), who is mostly known for his

1. http://en.wikipedia.org/wiki/E_(mathematical_constant)

work in fluid dynamics. Bernoulli became obsessed with the limit equation, and he actually calculated it out.

By the time Leibniz's calculus was published, *e* was well and truly entrenched, and we haven't been able to avoid it since.

Why the letter *e*? We don't really know. It was first used by Euler, but he didn't say why he chose that. It was probably an abbreviation for "exponential."

Does *e* Have a Meaning?

Does *e* mean anything? Or is it just an artifact—a number that's just a result of the way that numbers work?

That's more a question for philosophers than mathematicians. But I'm inclined to say that the number *e* is an artifact, but that the natural logarithm is deeply meaningful. The natural logarithm is full of amazing properties: it's the only logarithm that can be written with a closed-form series; it's got that wonderful interval property with the $1/x$ curve; it really is a deeply natural thing that expresses very important properties of the basic concepts of numbers. As a logarithm, the natural logarithm had to have some number as its base; it just happens that it works out to be the value *e*. But it's the logarithm that's the most meaningful, and you can calculate the natural logarithm without knowing the value of *e*.

7

φ: The Golden Ratio

Now we get to a number that really annoys me. I'm not a big fan of the golden ratio, also known as φ ("phi"). It's a number that has been adopted by all sorts of flakes and crazies, and there are alleged sightings of it in all sorts of strange places that are simply not real. For example, new-age pyramid worshipers claim that the great pyramids in Egypt have proportions that come from the golden ratio, but the simple truth is that they don't. Animal mystics claim that the ratio of drones to larvae in a beehive is approximately the golden ratio, but it isn't.

The thing is, the value of the golden ratio is (1 + sqrt(5))/2, or roughly 1.6. It's just a hair more than 1 ½. Sitting there, it's naturally *close to* lots of different things. If something is close to one and a half, it's close to the golden ratio. And because it has this reputation for being ubiquitous, people assume: oh, hey, look, it's the golden ratio!

For example, there was even a recent study that claimed that the *ideal* ratio of a woman's bust to her hip is related to the golden ratio. Why? Well, obviously, *everyone knows* that the golden ratio is the perfect aesthetic, and when they did a survey where men assessed the beauty of different pictures of women, if you squinted a bit when you looked at the results, the proportions of the women in the most highly rated pictures were close to the golden ratio. There's no real reason to believe this beyond the belief that the golden ratio is important, and there are many reasons *not* to believe it. (For example, the "ideal" figure for a woman has varied through history.)

But even if you discard all of this stuff, the golden ratio is an interesting number. My own personal reason for thinking it's cool is representational. In many different ways of writing numbers, the structure of the number becomes apparent in fascinating ways. For example, if you write the golden ratio as a continued fraction (which I explain in 11, *Continued Fractions*, on page 69), you get this:

$$\varphi = 1 + \cfrac{1}{1 + \cfrac{1}{1 + \cfrac{1}{1 + \cfrac{1}{\cdots}}}}$$

You could also write this as *[1;1,1,1,1...]*. And if you write it as a continued square root, it's this:

$$\varphi = 1 + \sqrt{1 + \sqrt{1 + \sqrt{1 + \sqrt{1 + \cdots}}}}$$

These different representations of φ aren't just pretty; they tell us something about the legitimately interesting properties of the number. The continued-fraction form of φ tells us that the reciprocal of φ is φ – 1. The continued root form means that $\varphi^2 = \varphi + 1$. Those are both different ways of explaining the fundamental geometric structure of φ, as we'll show in the next section.

What Is the Golden Ratio?

What is this golden ratio thing, anyway? It's the number that is a solution for the equation (a + b)/a = (a/b). In other words, given a rectangle where the ratio of the length of its sides is 1:φ, when you remove the largest possible square from it, you'll get another rectangle whose sides have the ratio φ:1. If you take the largest square from that, you'll get a rectangle whose sides have the ratio 1:φ. And so on. You can see an image of the basic idea in Figure 3, *The golden ratio*, on page 43.

Allegedly, the golden ratio is the ratio of the sides of a rectangle that produce the most aesthetically beautiful appearance. I'm not enough of a visual artist to judge that, so I've always just taken that on faith.

But the ratio does show up in many places in geometry. For example, if you draw a five-pointed star, the ratio of the

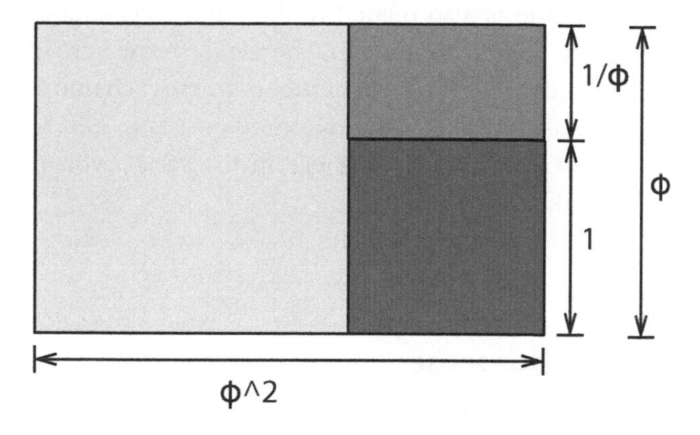

Figure 3—The golden ratio: *The golden ratio is the ratio of the sides of an ideal rectangle. When you remove the largest possible square from such a figure, you're left with another rectangle whose sides have same ratio.*

length of one of the point-to-point lines of the star to the length of the sides of the pentagon inside the star is φ:1, as shown in the next figure.

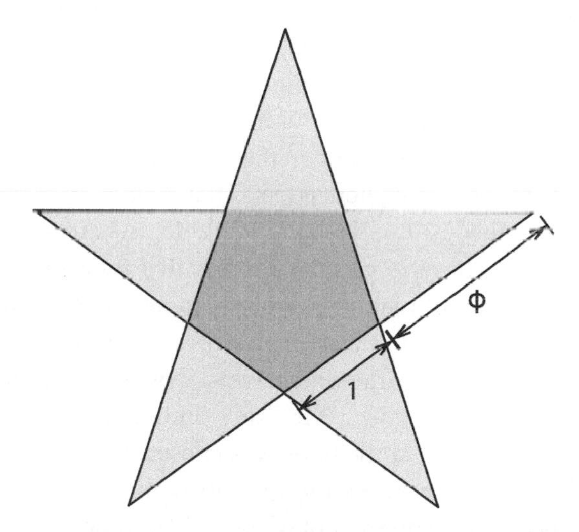

Figure 4—The golden ratio and the five-pointed star: *Another example of the golden ratio can be found in the ratio of the sides of the isosceles triangles surrounding the central pentagon of a five-pointed star.*

The golden ratio is also related to the Fibonacci sequence. In case you don't remember, here's a refresher: the Fibonacci sequence is the set of numbers where each number in the series is the sum of the two previous: 1, 1, 2, 3, 5, 8, 13, ... If *Fib(n)* is the *n*th number in the series, you can compute it like this:

$$\mathrm{Fib}\,(n) = \frac{\varphi^{n} - (1 - \varphi)^{n}}{\sqrt{5}}$$

Legendary Nonsense

There are a ton of stories about the history of the golden ratio. Most of them are utter bunk.

Many histories will tell you that the pyramids of Egypt are built on the golden ratio or that the proportions of various features of Greek temples were built to fit the golden ratio. But like so much else about the lore of the golden ratio, it's just apophenia—finding patterns where there are none.

Let's look at an example of the supposed connection between the pyramids and the golden ratio. If we look at the faces of the Great Pyramid of Khufu, the relation to the golden ratio is just a rough approximation: the golden ratio is roughly 1.62, and the ratio of the length of the sides of the Great Pyramid is 1.57. The Great Pyramid is famous for the precision of its construction, but on the scale of the pyramid, that error corresponds to about 6 feet. That kind of error doesn't fit the mythology. But so many mythologists are so certain that the golden ratio *must* be a factor that they struggle to find some way of making it work. People will go to amazingly great lengths to satisfy their expectation. For example, one author who's written extensively about the mythologies of the great pyramids, a Mr. Tony Smith, argues that the ratio of the length of the faces was chosen so that the angle at which the side meets the ground is the arc-sine of the reciprocal of the square root of the golden ratio.[1]

We do know that the golden ratio was identified by someone from the cult of Pythagoras, quite possibly Hippasus, our poor drowned friend from the history of irrational numbers.

---

1. http://www.tony5m17h.net/Gpyr.html

The golden ratio was well known among the ancient Greeks. Euclid wrote about it in his *Elements*, and Plato wrote about it in his philosophical works. In fact, I would argue that Plato is the initiator of all of the bizarre, flaky metaphysical gibberish frequently attached to the golden ratio. He believed that the world was made of up four elements, each of which was formed from perfect polyhedra. These perfect polyhedra were, according to Plato, built from triangles that were themselves formed according to perfect ratios—foremost among them was the golden ratio. To Plato, the golden ratio was one of the most fundamental parts of the universe.

The reason the golden ratio is called φ is actually a reference to the Greek sculptor Phidias (cc. 490–430 BC), who used it in his compositions. Written in Greek, his name started with the letter φ.

After the Greeks, there wasn't a lot of interest in φ until the sixteenth century, when a monk named Pacioli (1445–1517), known for his studies of both art and mathematics, wrote a text called *The Divine Proportion*, which discussed the golden ratio and its uses in architecture and art. Da Vinci was fascinated by the golden ratio as he studied Pacioli's text, and as a result, it ended up playing a major role in many of his sketches and paintings. In particular, his infamous "Vitruvian Man" (shown in Figure 5, *DaVinci's Vitruvian Man*, on page 46) is his illustration of how the human body supposedly embodies the divine proportion.

Of course, once Da Vinci embraced it, artists and architects all over Europe immediately jumped on the bandwagon, and it's pretty much continued to be used by artists and architects all the way to the present.

Where It Really Lives

As I've been explaining, people constantly see the golden ratio where it isn't. But still, it is a real thing, and it does manage to show up in some astonishingly odd places.

Most of the real appearances of the golden ratio are related to the Fibonacci sequence. Because the Fibonacci sequence and the golden ratio are deeply connected, wherever the Fibonacci sequence shows up, you can find the golden ratio.

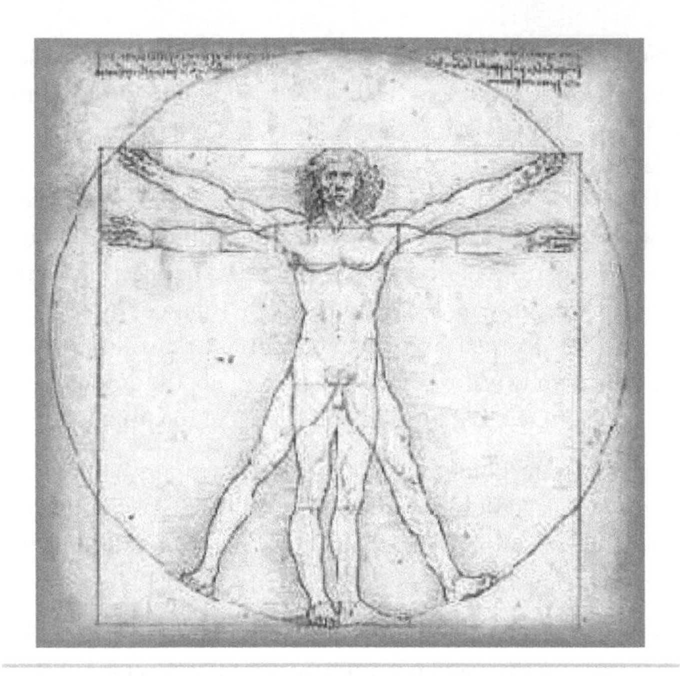

Figure 5—DaVinci's *Vitruvian Man*: *DaVinci believed that the human form embodied the golden ratio.*

For example, the basic scale used by Western music is built on the Fibonacci sequence, and the chord structure of most tonal music has numerous examples of the golden ratio between the tones that make up the chords. Several musicians have taken advantage of this pattern. My favorite example of this comes from the great twentieth-century composer Béla Bartók (1881–1945), who used it as a fundamental construct in some of his works, most wonderfully as a portion of the twelve-part fugue in his *Music for Strings, Percussion, and Celesta*—to my knowledge one of the only twelve-part fugues in the canon of European music.

For fun, you can build a number system called phinary that is based on the golden ratio. It's a strange number system that has some amusing properties. In phinary, because of the particular structure of phi as an irrational number, every rational number has a *non-terminating* representation: that means that every rational number looks irrational in phinary. Is phinary useful for anything? Not particularly, but it's pretty neat anyway!

8

i: The Imaginary Number

Possibly the most interesting strange number is the sadly maligned *i*, the square root of −1, also known as the "imaginary" number. Where'd this strange thing come from? Is it real (not in the sense of real numbers, but in the sense of representing something real and meaningful in our world)? What's it good for?

The Origin of *i*

The number *i* has its "roots" in the work of early Arabic mathematicians, the same people who first really understood the number. But they weren't quite as good with *i* as they were with 0; they didn't really get it. They had some concept of the roots of a cubic equation, where sometimes the tricks for finding the roots just didn't work. They knew there was something going on, some way that the equation needed to have roots, but just what that really meant, they didn't get.

Things stayed that way for quite a while. Mathematicians studying algebra knew that there was something missing in a variety of mathematical concepts, but no one understood how to solve the problem. What we know as algebra developed for many years, and various scholars, like the Greeks, encountered them in various ways when things didn't work, but no one really grasped the idea that algebra required numbers that were more than just points on a one-dimensional number line.

The first real step toward *i* was in Italy, over a thousand years after the Greeks. During the sixteenth century, mathematicians were searching for solutions to cubic equations, the same things that the early Arabian scholars had tried to find. But finding solutions for cubic equations, even when those equations did have real solutions, sometimes required working with the square root of –1 along the way.

The first real description of *i* came from a mathematician named Rafael Bombelli (1526–1572), who was one of the mathematicians trying to find solutions for cubic equations. Bombelli recognized that you needed to use a value for the square root of –1 in some of the steps to reach a solution, but he didn't really think that *i* was something real or meaningful in numbers; he just viewed it as a peculiar but useful artifact for solving cubic equations.

i got its unfortunate misnomer, the "imaginary number," as a result of a diatribe by the famous mathematician/philosopher René Descartes (1596–1650). Descartes was disgusted by the concept of *i*, believing that it was a phony artifact of sloppy algebra. He did not accept that it had any meaning at all: thus he termed it an "imaginary" number as part of an attempt to discredit the concept.

Complex numbers built using *i* finally came into wide acceptance as a result of the work of Leonhard Euler (1707–1783) in the eighteenth century. Euler was probably the first to truly fully comprehend the complex number system created by the existence of *i*. And working with that, he discovered one of the most fascinating and bizarre mathematical discoveries ever, known as Euler's equation. I have no idea how many years it's been since I was first exposed to this, and I still have a hard time wrapping my head around why it's true.

$$e^{i\theta} = \cos\theta + i\sin\theta$$

And here's what that really means:

$$e^{i\pi} = -1$$

That's just astonishing. The fact that there is such a close relationship between i, π, and e is just shocking on the face of it.

What *i* Does

Once the reality of *i* as a number was accepted, mathematics was changed irrevocably. Instead of the numbers described by algebraic equations being points on a line, suddenly they become points *on a plane*. Algebraic numbers are really two-dimensional; and just like the integer *1* is the unit distance on the axis of the real numbers, *i* is the unit distance on the axis of the imaginary numbers. As a result numbers in general become what we call *complex*: they have two components, defining their position relative to those two axes. We generally write them as $a + bi$, where a is the real component and b is the imaginary component. You can see in the following figure what a complex number as a two-dimensional value means.

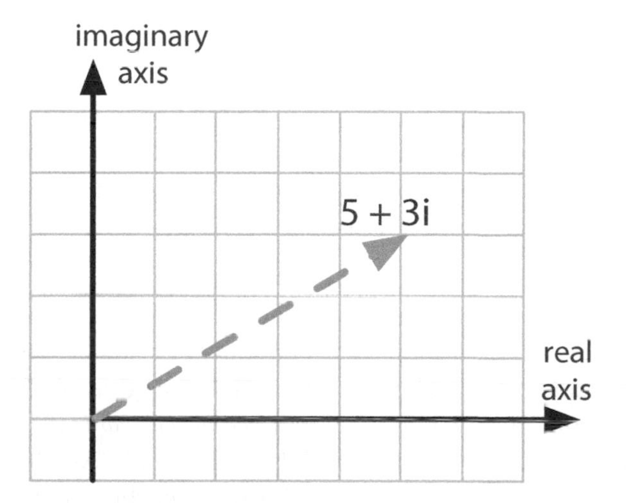

Figure 6—Complex number as a point on a 2D plane: *A complex number* a + bi *can be represented graphically as a point on a two-dimensional plane, where* a *is its position along the real axis, and* b *is its position along the imaginary axis.*

The addition of i and the resulting addition of complex numbers is a wonderful thing mathematically. It means that every polynomial equation has roots. In particular, a polynomial equation in x with a maximum exponent n will always have exactly n complex roots.

But that's just an effect of what's really going on. The real numbers are not closed algebraically under multiplication and addition. With the addition of i, multiplicative algebra becomes closed, which means that every operation and every expression in algebra becomes meaningful. Nothing escapes the system of the complex numbers.

Of course, it's not all joy and happiness once we go from real to complex numbers. Complex numbers aren't ordered. There is no less-than (<) comparison for complex numbers. The ability to do meaningful inequalities evaporates when complex numbers enter the system.

What *i* Means

But what do complex numbers mean in the real world? Do they really represent actual phenomena, or are they just a mathematical abstraction?

The answer is that they're very real, as any scientist or engineer will tell you. There's one standard example that everyone uses because it's so perfect: Take the electrical outlet that powers your lights, your computer. It provides an alternating current. What does that mean?

Well, the voltage—which (to oversimplify) can be viewed as the amount of force pushing the current— is complex. In fact, if your outlet supplies a voltage of 110 volts AC at 60 hz (the standard in the US), that means the voltage is a number of magnitude 110. In Figure 7, *Alternating current in the complex plane: electric vs. magnetic fields*, on page 51, I've plotted the real voltage on a graph with time on the x-axis and voltage on the y-axis. That's a sine wave. Along with that, I plotted the strength of the magnetic field, which is a sine wave out of phase with the electric field by 90 degrees, so that when the magnetic field is at its maximum, the voltage is 0.

If you looked at just the voltage curve, you'd get the wrong impression because it implies that the power is turning on and off really quickly. It isn't: there's a certain amount of power being transmitted in the alternating current, and that amount of power is constant over time. The power is being expressed in different ways. From our perspective as

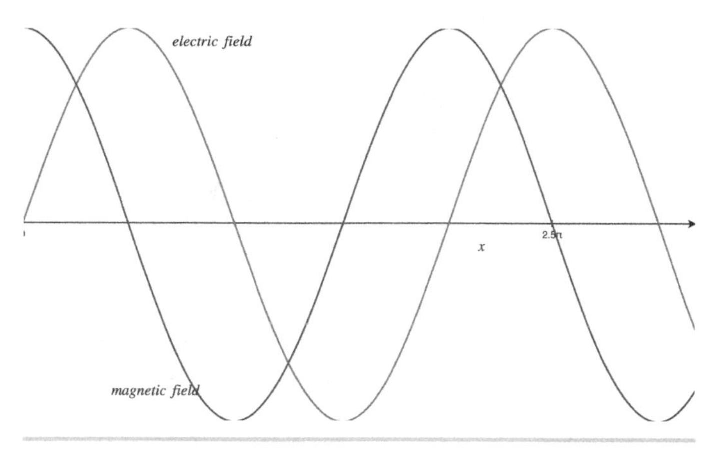

Figure 7—Alternating current in the complex plane: electric vs. magnetic fields

electricity users, we can usually treat it as turning on and off, because we're only using the voltage part of the power system. In reality, though, it's not really turning on and off: it's a dynamic system, a system in constant motion, and we're just looking at one slice of that motion.

The vector representing the power being transmitted is a fixed-size vector. It's rotating through the complex number plane, as I try to show in Figure 8, *Rotation through the complex plane*, on page 52. When it's rotated entirely into the imaginary plane, the energy is expressed completely as a magnetic field. When it's rotated entirely into the real plane, the energy is expressed completely as an electric field, which we measure as the voltage. The power vector isn't shrinking and growing; it's *rotating*.

The relationship between the electric and magnetic fields in AC electricity is really typical of how *i* applies in the real world: it's a critical part of fundamental relationships in dynamic systems with related but orthogonal aspects, where it often represents a kind of rotation or projection of a moving system in an additional dimension.

You can see another example of the same basic idea in computer speech processing. We analyze sounds using something called *the Fourier transform*. To be able to translate sound into words, one of the tricks engineers use is to decompose a complex waveform (like the sound of a human

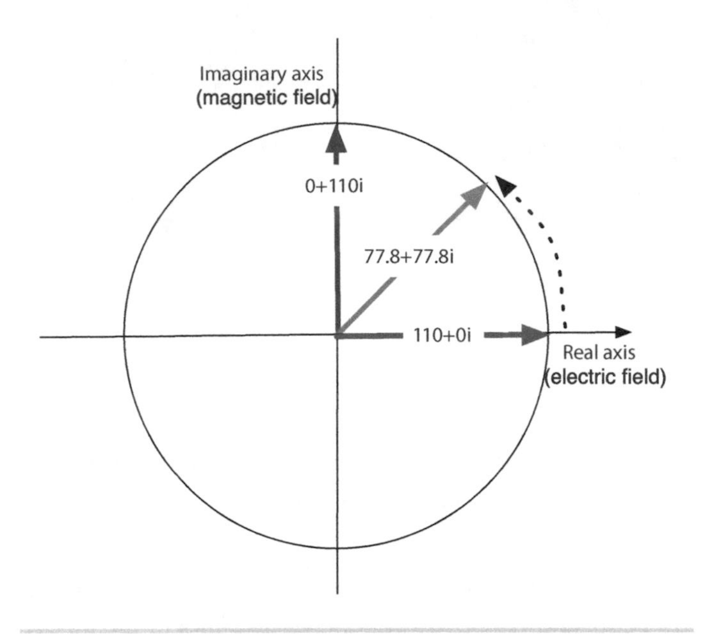

Figure 8—Rotation through the complex plane

voice) into a collection of basic sine waves, where the sum of the sine waves equal the wave at a given point in time The process by which we do that decomposition is intimately tied with complex numbers: the Fourier transform and all of the analyses and transformations built on it are dependent on the reality of complex numbers (and in particular on the magnificent Euler equation that we introduced at the start of this section).

Part III

Writing Numbers

Numbers are so familiar to us that we often forget about their beauty and their mystery. We think that the way we commonly look at numbers is the only way to look at them and that the way we write them is the only correct way to do so.

But that isn't always the case. As mathematics has evolved, its practitioners have invented an amazing number of ways to write numbers. Some of these systems of notation are worse than the ones we use today, and some are just, well, *different*. In this part of the book, we're going to look at two ancient systems of notation and a modern one that continues to intrigue today's mathematicians.

We'll look first at how the Romans wrote numbers and see how our number system really does make a dramatic difference in how easily we can do math by looking at how difficult it is to do using roman numerals.

We'll look at how the Egyptians wrote fractions and see how the aesthetics of numbers have changed.

And we'll look at an example of how even today mathematicians have continued to invent new ways of writing numbers for specific purposes by looking at a special kind of fraction, called a *continued fraction*.

9

Roman Numerals

We normally write numbers in a notation that's called arabic numbers, because it came to the European culture via Arabian mathematicians. But before that, in Western culture, numbers were written in the Roman style. In fact, in a lot of formal documents, we *still* use roman numerals. If you look at the preface to most textbooks, the pages are numbered with roman numerals. Every day on my way to work, I walk past buildings in Manhattan that have cornerstones recording the year the building was constructed, and every one uses roman numerals. In the closing titles of most movies, the copyright date is written in roman numerals.

Even though I've probably seen roman numerals every day of my life, I've always been perplexed by them. Why would anyone come up with something so strange as a way of writing numbers? And given that they're so damned weird, hard to read, and hard to work with, why do we still use them for so many things today?

A Positional System

I expect most people already know this, but I'll go through an explanation of how roman numerals work. The roman numeral system is nonpositional, meaning that a numeric symbol represents a particular value no matter where it sits. This is very different from our decimal arabic notation, which is positional. In our common notation, the 3 in "32" represents thirty, but the 3 in "357" represents three hundred. In roman numerals, that's not true. An "X" always represents ten, no matter where in the number it sits.

The basic scheme of roman numerals is to assign fixed numeric values to letters.

- "I" stands for 1.
- "V" stands for 5.
- "X" stands for 10.
- "L" stands for 50.
- "C" stands for 100.
- "D" stands for 500.
- "M" stands for 1000.

The standard set of roman numerals doesn't include any symbols for representing numbers larger than 1000. In the middle ages, monks that used roman numerals in manuscripts added a notation for larger numbers; adding a horizontal line over the top of a numeral means that numeral multiplied by 1000, so that a V with a horizontal line floating over it represents 5000, an X with a line floating over it represents 10,000, and so forth. But that's a later addition, which isn't part of the original roman numeral system. Even with that addition, it's difficult to clearly write larger numbers.

The symbols in roman numerals are combined in a bizarre way. Take a roman numeral like I or X. When two or more appear in a group—such as III or XXX—they are added together. Thus, III represents 3, and XXX represents 30. When the number represented by a numeral is *smaller* than the one to its right, it is *subtracted* from its successor. But if the order is reversed with the smaller number coming *after* the larger one, then it is *added* to the number to its left.

The notation for a number is anchored by the *largest* roman numeral symbol used in that number. In general (though not always), you do not precede a symbol by anything smaller than one-tenth its value. So you wouldn't write IC (100 – 1) for 99; you'd write XCIX (100 – 10 + 10 – 1).

Let's look at a few other examples.

- *IV* = *4*: V = *5*, I = *1*. I precedes V so it's subtracted, so *IV* = *5* – *1*.

- *VI* = *6*: V = *5*, I = *1*. I follows V so it's added, so *VI* = *5* + *1* = *6*.

- *XVI = 16*: *X = 10, V = 5, I = 1*. *VI* is a number starting with a symbol whose value is smaller than *X*, so we take its value and add it. Since *VI = 6*, then *XVI = 10 + 6 = 16*.
- *XCIX = 99: C = 100*. The X preceding the C is subtracted, so *XC = 90*. Then the IX following it is added. X is ten, preceded by I, so *IX = 9*. So *XCIX = 99*.
- *MCMXCIX = 1999*: *M = 1000*. CM is *1000 – 100 = 900*, so *MCM – 1900*. *C = 100, XC = 90*. *IX = 9*.

For some reason, 4 is sometimes written IV and sometimes IIII (there are a number of theories why, which I'll discuss later).

What about 0? Is there a roman numeral for 0? Sort of. It wasn't part of the original system and was never used by the Romans themselves. But during the Middle Ages, monks using roman numerals used *N*, for *nullae*, to represent 0. But it wasn't the positional 0 of arabic numbers; it was just a roman numeral used to fill into the astronomical tables used to compute the date of Easter, rather than leaving the column blank.

Where Did This Mess Come From?

The main theory about the origin of roman numerals is that they were invented by shepherds, who counted their flocks by marking notches on their staffs. The system that became roman numerals started off as just these notches on a staff, not letters at all.

When counting their sheep, a shepherd would mark four notches, one at a time for the first four; and then on the fifth one, they would cut a diagonal notch, in pretty much the same way that we commonly write four lines and then a diagonal strike-through for five when tallying something. But instead of striking through the preceding notches, they just used the diagonal to turn a "/" notch into "V." Every tenth notch was marked by a strike-through so it looked like an "X." Every tenth V got an extra overlapping notch, so it looked sort of like the Greek letter *psi*, and every tenth X got an extra overlapping notch, so it looked like an X with a vertical line through the center.

In this system, if you had eight sheep, you would write IIIIVIII. But the leading IIIIs are not really needed. So you could just use VIII instead, which became important when you wanted to write a big number.

When the Romans made this system part of their written language, the simple notches became I and V, the strike-through became X, and the *psi*-like thing became L. Beyond that they started using mnemonics: the symbols C, D, and M were derived from the Latin words for 100, 500, and 1000.

The prefix-subtraction stuff came as it transitioned to writing. The problem with an ordinal system like this is that it involves a lot of repeated characters, which are very difficult for people to read correctly. Keeping the number of repetitions small reduces the number of errors that people make reading the numbers. It's more compact to write IX than VIIII, and it's a lot easier to read because of fewer repetitions. So scribes started using the prefix-subtraction form.

Arithmetic Is Easy (But an Abacus Is Easier)

Looking at the roman numerals, it looks like doing arithmetic in that format will be a nightmare. But basic arithmetic—addition and subtraction—is pretty easy. Addition and subtraction are simple, and it's obvious why they work. On the other hand, multiplication with roman numerals is difficult, and division is close to impossible. It's worth noting that while scholars did teach arithmetic this way, most everyday calculations were done in Roman times with a Roman abacus, which works much more like our modern number system does.

To add two roman numerals, you do this:

1. Convert any subtractive prefixes to additive suffixes. So, for example, IX would be rewritten to VIIII.
2. Concatenate (or link together) the two numbers to add.
3. Sort the letters large to small.
4. Do internal sums (for example, replace IIIII with V)
5. Convert back to subtractive prefixes.

Example: 123 + 69. In roman numerals, that's CXXIII + LXIX.

- CXXIII has no subtractive prefixes. LXIX becomes LXVIIII.
- Concatenate: CXXIIILXVIIII.
- Sort: CLXXXVIIIIIII.
- Internal sum: reduce the IIIIII to VII, giving CLXXXVVII; then reduce the VV to X, resulting in CLXXXXII.
- Switch to subtractive prefix: XXXX = XL, giving CLXLII. LXL = XC, giving CXCII, or 192.

Subtraction isn't any harder than addition. To subtract *A* – *B*, you would take these steps:

1. Convert subtractive prefixes to additive suffixes.
2. Eliminate any common symbols that appear in both *A* and *B*.
3. For the largest remaining symbol in *B*, take the first symbol in *A* larger than it and expand it into repetitions of the next largest unit. Then go back to step 2 until there's nothing left. (For example, to subtract XX from L, you'd first expand the L to XXXXX.)
4. Convert back to subtractive prefixes.

Example: 192 – 69, or CXCII – LXIX.

- Remove prefixes: CLXXXXII – LXVIIII.
- Remove common symbols: CXXX – VII.
- Expand an X in CXXX: CXXVIIIII – VII.
- Remove common symbols: CXXIII = 123.

Doing multiplication with roman numerals is neither easy nor obvious. Both figuring out just how to make it work and actually doing it are difficult. You can do the trivial thing, which is repeated addition. But it should be pretty obvious that that's not practical for large numbers. The trick that the Romans used was clever, actually! It's basically a strange version of binary multiplication. To make it work, you need to be able to do addition and to divide by two, but both are pretty easy things to do. So here goes:

1. Given $A \times B$, you create two columns, and write A in the left column, and B in the right.
2. Divide the number in the left column by 2, discarding the remainder. Write the result down in the next row of the left column.
3. Multiply the number in the right column by 2. Write that answer down in the right column next to the result from step 1.
4. Repeat step 1 through 3 until the value in the left column is 1.
5. Go down the table and cross out every row where the number in the left column is even.
6. Add up the remaining values in the right column.

Let's look at an example: 21 × 17, or XXI × XVII in roman numerals.

We build the table:

Left	Right
XXI (21)	XVII (17)
X (10)	XXXIV (34)
V (5)	LXVIII (68)
II (2)	CXXXVI (136)
I (1)	CCLXXII (272)

Then strike out the rows where the left side is even:

Left	Right
XXI (21)	XVII (17)
V (5)	LXVIII (68)
I (1)	CCLXXII (272)

Now add the right column: XVII + LXVIII + CCLXXII = CCLLXXXXVVIIIIIII = CCCXXXXXVII = CCCLVII = 357.

Why does it work? It's binary arithmetic. In binary arithmetic, to multiply A by B, you start with 0 for the result, and then for each digit d_n of A, if $d_n = 1$, then add B with n 0s appended to the pending result.

Dividing by two gives you the binary digit of *A* for each position: if it's odd, then the bit in that position is 1; if it's even, the bit in that position is 0. Multiplying by two on the right is giving you the results of appending the zeros in binary—for the nth digit, you've multiplied by two *n* times.

Division is the biggest problem in roman numerals. There is no good trick that works in general. Division really comes down to making a guess, doing multiplication to see if it's right, and then adjusting the guess. The only thing you can do to simplify is variations on finding a common factor of both numbers that's easy to factor out. For example, if both numbers are even, you can divide each of them by two before starting the guesswork and testing. It's also fairly easy to recognize when both numbers are multiples of 5 or 10 and to do the division by 5 or 10 on both numbers. But beyond that, you take a guess, do the multiplication, subtract, repeat.

Blame Tradition

We use roman numerals for historic reasons. Until quite recently, scholars in western culture did most of their work in Latin. For example, Isaac Newton (1643–1727) wrote his famous monograph, *Philosophiae Naturalis Principia Mathematica,* in Latin in the seventeenth century because all scholarly work at the time was published in Latin.

Using a historic language out of tradition, it also made sense to use their traditional number notation. This has continued into the present in many places entirely out of tradition. For example, modern architects still put dates on cornerstones using roman numerals. While a modern geek like me would argue that it doesn't make sense to use such an impractical notation, tradition is a powerful force, and tradition dominated.

There's no practical reason for us to continue to use roman numerals. It's just a matter of tradition. But even with the question of *why* answered, there's a lot of strangeness around our use of roman numerals, which can't just be explained by tradition.

One very common question is "Why does a clock use IIII instead of IV?"

The answer is unclear. There are a bunch of different theories, and no one is really entirely sure which one is correct. The most common ones, along with my opinions of them, include these:

- *I and V are the first letters of the name of the god Jupiter in Latin.* Probably bogus, but still widely cited. Romans had no issue with writing down the name Jupiter. Worrying about writing the name of a deity is an issue that comes from the ten commandments in Judaism and Christianity.

- *I and V could be the first letters of the name "Jehovah" in Latin.* Better than the Jupiter thing, since early Christians did follow the Jewish practice of not writing God's name. I'm still skeptical, but at least it's historically consistent.

- *IIII is more symmetric with VIII on the clock face.* Pretty likely—our style of clock dates back to the artisans that designed the first clock faces. Artists and craftsmen are often extremely obsessed with aesthetics and balance, and it's true that IIII and VIII really *do* look better than IV and VIII.

- *IIII allows clock makers to use fewer molds to make the numbers for the clock face.* Probably not; there's not a big enough difference.

- *The king of France liked the way that IIII looked better than IV.* Once again, I'm very skeptical. People love to blame things on despicable aristocrats. But clock faces don't really historically date back to France, and there's no contemporary documentation to back this up.

- *Coincidence.* Technically, IIII is as correct as IV. So someone who started making clocks just happened to be someone who used IIII instead of IV. In fact, the Romans themselves generally preferred IIII; it's more common in historical documents. And that last fact is, quite probably, the real reason.

And before I close this section, my favorite question: What's the silliest thing that anyone's done with roman numerals?

There's a delightfully ridiculous programming language, designed as an elaborate joke, called INTERCAL (which stands for "Language With No Pronounceable Acronym").[1]

INTERCAL as it was originally designed was amazingly horrible. But then a bunch of enthusiasts got hold of it and decided to make it worse. The original INTERCAL didn't really have a way to do input or output. So to make INTERCAL both more complete and more horrible, its inventors decided to add input and output but chose to use their own, custom variant of roman numerals. The result could give any sane programmer screaming nightmares. In INTERCAL, they follow the rule we mentioned about using a horizontal bar floating over a number to multiply by 1000. But when INTERCAL was invented, they were programming for a teletype, which didn't have a character for a letter with a bar over it. So they defined a variant of roman numerals that includes backspace characters. 5000 was printed as "V<backspace>-."

Looking at this, one thing should be clear: we're *really* lucky that arabic numerals took the place of roman numerals. Everything is harder with roman numerals. And if we still really used roman numerals, we might all be stuck programming in INTERCAL.

---

1. You can find out more about the horror that is INTERCAL at the INTERCAL resources site: http://catb.org/esr/intercal/.

10

Egyptian Fractions

As math has evolved, so have people's views of what is aesthetic in math and what is not. The ancient Greeks (from whom we got many of our mathematical notions) thought, for example, that as a matter of elegance, the only way to write fractions was as *unit* fractions—fractions whose numerator is 1. A fraction that was written with a numerator larger than one was considered *wrong*. Even today, many math books use the term *vulgar fraction* to refer to non-unit fractions.

Obviously, there are fractions other than the unit fractions. A unit fraction like $\frac{1}{3}$ is a reasonable quantity; but so is $\frac{2}{3}$. So how did the Greeks handle these quantities? They represented them in a form that we refer to today as an *Egyptian fraction*. An Egyptian fraction is expressed as the sum of a finite set of unit fractions. For example, instead of writing the vulgar fraction $\frac{2}{3}$, the Greeks would write "$\frac{1}{2} + \frac{1}{6}$."

A 4000-Year-Old Math Exam

We don't know that much about the origins of Egyptian fractions. What we do know is that the earliest written record of their use is in an Egyptian scroll from roughly the eighteenth century BC, which is why they're known as Egyptian fractions.

That scroll, known as the *Rhind Papyrus*, is one of the most fascinating documents in the entire history of mathematics. It appears to be something along the lines of a textbook of Egyptian mathematics! It's a set of what look like exam questions along with fully worked answers. The scroll includes tables of fractions written in unit-fraction sum form, as well as numerous algebra (in roughly the form we use today!) and geometry problems. From the wording of the scroll, it's strongly implied that the author is recording techniques well-known to the mathematicians of the day but kept secret from the masses. (What we would call mathematicians were part of the priestly class in Egypt, usually temple scribes. Subjects such as advanced math were considered a sort of sacred mystery reserved for those in the temples.)

So we don't really know when Egyptian fractions were invented or by whom. But from the time of the Egyptians through the empires of the Greeks and Romans, they continued to be considered the correct mathematical notation for fractions.

As I said in the introduction, vulgar fractions were considered ugly at best and incorrect at worst all the way into the Middle Ages. Fibonacci defined what is still pretty much the canonical algorithm for computing the Egyptian fraction form of a rational number.

Not too long after Fibonacci, the obsession with avoiding vulgar fractions declined. But they've stayed around both because of the historical documents that use them and because they're useful as a way of looking at certain problems in number theory (not to mention as a foundation for a lot of nifty mathematical puzzles).

Fibonacci's Greedy Algorithm

The primary algorithm for computing the Egyptian fraction form is a classic example of what computer-science geeks like me call a *greedy algorithm*. The greedy algorithm doesn't always generate the shortest possible Egyptian fraction form, but it is guaranteed to terminate with a finite (if ugly) sequence.

The basic idea of the algorithm is this: given a vulgar fraction x/y, its Egyptian form can be computed like this:

$$e(\frac{x}{y}) = \frac{1}{\lceil y/x \rceil} + e(r)\text{, where } r = \frac{-y \bmod x}{y \times \lceil y/x \rceil}$$

Or in a slightly more useful form, here's the same algorithm written as a Haskell program that returns a list of unit fractions. (For non-Haskell folks out there, x%y is a Haskell-type constructor that creates the fraction x/y; and x:y creates a list with head x and tail y.)

```
egypt :: Rational -> [Rational]
egypt 0 = []
egypt fraction =
  (1%denom):(remainders) where
    x = numerator fraction
    y = denominator fraction
    denom = ceiling (y%x)
    remx = (-y) `mod` x
    remy = y*denom
    remainders = egypt (remx%remy)
```

And for fun, here's the reverse process, converting from the Egyptian form to the vulgar:

```
vulgar :: [Rational] -> Rational
vulgar r = foldl (+) 0 r
```

To get a sense of what Egyptian fractions look like and how complex they can get, let's look at a few examples.

Example: Egyptian fractions

- 4/5 = 1/2 + 1/4 + 1/20
- 9/31 = 1/4 + 1/25 + 1/3100
- 21/50 = 1/3 + 1/12 + 1/300
- 1023/1024 = 1/2 + 1/3 + 1/7 + 1/44 + 1/9462 + 1/373029888

As you can see, the Fibonacci algorithm for Egyptian fractions can generate some really ugly terms. It often generates sequences of fractions that are longer than necessary and that include ridiculously large and awkward denominators. For example, that last fraction can be written more clearly as (1/2 + 1/4 + 1/8 + 1/16 + 1/64 + 1/128 + 1/256 + 1/512 + 1/1024). One of the canonical examples of weakness of the Fibonacci algorithm is 5/121 = 1/25 + 1/757 + 1/763309 + 1/873960180913 + 1/1527612795642093418846225, which can be written much more simply as 1/33 + 1/121 + 1/363.

The problem, though, is that the Fibonacci algorithm is the most widely known and easiest to understand of the methods for computing Egyptian fractions.[1] While we can compute them, we don't know of any particularly good or efficient ways of computing the minimum-length forms of Egyptian fractions. In fact, we don't even know what the complexity bounds of computing a minimal Egyptian fraction are.

Sometimes Aesthetics Trumps Practicality

What I find particularly interesting about Egyptian fractions is how long they've lasted given how difficult it is to work with them. Adding Egyptian fractions is difficult; multiplying one by an integer is a pain, but multiplying two of them is absolutely insane. From a purely practical standpoint, they seem downright ridiculous. As early as AD 150, they were roundly criticized by Ptolemy himself! And yet they were the dominant way that fractions were written for close to three thousand years. The aesthetics of unit fractions overwhelmed the practicality of tractable arithmetic.

There are a bunch of interesting, open problems involving Egyptian fractions. I'll just leave you with one fun example: Paul Erdös, the great Hungarian mathematician, tried to prove that for any fraction $4/n$, there was an Egyptian fraction containing exactly three terms. Doing brute-force tests, it's been shown to be true for every number n smaller than 1014, but no one has figured out how to prove it.

1. David Eppstein has a website with a collection of other Egyptian fraction algorithms, and implementations of many of them, at http://www.ics.uci.edu/~eppstein/numth/egypt/.

11

Continued Fractions

The way that we write numbers that aren't integers is annoying and frustrating.

We've got two choices: we can write them as fractions or we can write them as decimals. But both have serious problems. There are some numbers that we can write easily as fractions, like 1/3 or 4/7. That's great. But there are some numbers that you just can't write as a fraction, like π.

For numbers like π, there are fractions that are *close*. 22/7 is a common approximation for π. But fractions are terrible for that: if you need to be a bit more precise than that, you can't change it a little bit; you need to come up with a totally different fraction. That's one of the reasons that decimals are great. We can approximate a number like π using a decimal form like 3.14. If you need it to be a bit more precise, you can add a digit, like 3.141, then 3.1415, and so on.

But decimals have their own problems. Lots of numbers that we could have written perfectly as fractions, we can't ever write exactly as a decimal. We can write 1/3, but as a decimal? It's 0.33333333, repeating forever.

They're both great representations in their way. But they both fail in their own ways.

Can we do better? Yes. There's another way of writing fractions that keeps all of the good things about fractions and also gives this new kind of fraction all of the benefits of writing decimals. It's called *continued fractions*.

Continued Fractions

A continued fraction is a very neat thing. Here's the idea: take a number where you don't know its fractional form. Pick the nearest simple fraction *1/n* that's just a little bit too large. If you were looking at, say, 0.4, you'd take 1/2, because it's a bit bigger. That gives you a first approximation of the number as a fraction. But it's a little bit too big. If the value of a fraction is a little bit larger than you want it to be, that means that the denominator of that fraction is a little bit too small, and to fix it you need to add a correction to the denominator to make it a little bit bigger. A continued fraction works on that basic principle. Just keep adjusting the denominator; you approximate the correction to the denominator by adding a fraction to it that's just a little bit too big, and then you add a correction to the correction.

Let's look at an example:

Example: Express 2.3456 as a continued fraction.

1. It's close to 2. So we start with 2 + (0.3456).
2. Now we start approximating the fraction. We take the reciprocal of 0.3456 and take the integer part of the result: 1/0.3456 rounded down is 2. So we make it 2 + 1/2; and we know that the denominator is off by about 0.893518518518518.
3. We take the reciprocal again and get 1, and it's off by about 0.1191709844559592.
4. We take the reciprocal again and get 8, and it's off by about 0.3913043478260416.
5. Next we get 2 and it's off by about 5/9.
6. If we keep going, we'll get 1, 1, and 4, and then we'll have no remaining error.

And there you are. As a continued fraction, 2.3456 looks like this:

$$2.3456 = 2\cfrac{1}{2+\cfrac{1}{1+\cfrac{1}{8+\cfrac{1}{2+\cfrac{1}{1+\cfrac{1}{1+\frac{1}{4}}}}}}}$$

As a shorthand, continued fractions are normally written as a list notation enclosed by square brackets: the integer part comes first, followed by a semicolon, and then a comma-separated list of the denominators of the fractions in the series. So our continued fraction for 2.3456 would be written [2; 2, 1, 8, 2, 1, 1, 4]. When we write it in this form, as a sequence of terms, we call each term of the sequence a *convergent*.

There's a very cool visual way of understanding that algorithm. I'm not going to show the procedure for 2.3456, because it would be hard to draw the complete diagram for it in a legible way. So instead, we'll look at a simpler number. Let's write 9/16 as a continued fraction.

We start by drawing a grid for the fraction. The number of columns in the grid is the value of the denominator; the number of rows in the grid is the numerator. For 9/16, that means that our grid is 16 across by 9 down, like the one drawn in the following figure.

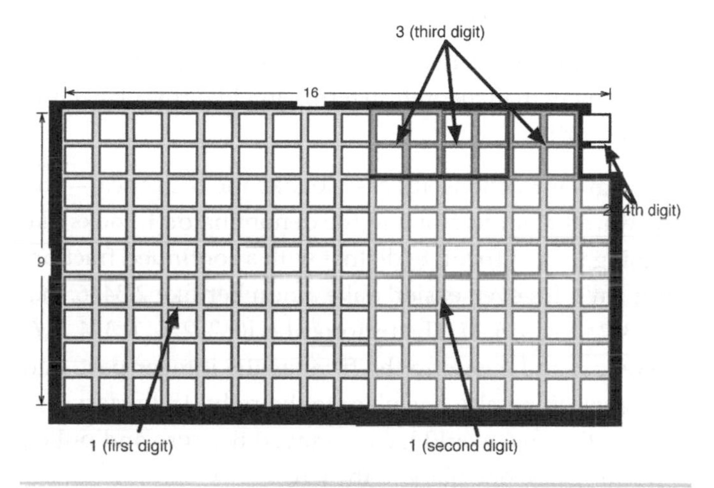

Figure 9—Computing a continued fraction graphically: *The most commonly used algorithm for computing a continued fraction can be visualized by extracting the largest possible squares from a grid whose height and width correspond to the numerator and denominator of a fraction. When 1 by 1 squares are all that remain, the fraction is complete.*

Now we draw the largest square we can on that grid. The number of squares of that size that we can draw is the first term of the continued fraction. For 9/16, we can draw only one 9-by-9 square, which is the largest square possible, so our first convergent is 1. Once we've marked off that square, we're left with a 7-by-9 rectangle.

Now we repeat the previous step: once again we draw the largest square we can. That's a 7-by-7 square and we can only draw one. So the second convergent is 1, and we're left with a 7-by-2 rectangle.

When we repeat the previous step, the largest square we can draw is one that's 2 by 2, but this time we can draw three of them. That means the next convergent is a 3. We're left with a 1-by-2 rectangle, in which it's possible to draw two 1-by-1 squares. So the last convergent is a 2.

No squares remain. The continued fraction for 9/16 is 1/(1+1/(1+3/(1+1/2))), or [0; 1, 1, 3, 2].

Cleaner, Clearer, and Just Plain Fun

Continued fractions really are interesting buggers. They're interesting both in theoretical terms and in terms of just fun, odd properties.

With regular fractions, it's easy to take the reciprocal—you just swap the numerator and the denominator. It looks like it's going to be difficult to do that with a continued fraction, but it isn't. It's even easier! Take a number like 2.3456, aka *[2; 2, 3, 1, 3, 4, 5, 6, 4]*. The reciprocal is [0; 2, 2, 3, 1, 3, 4, 5, 6, 4]. We just add a zero to the front as the integer part, and push everything else one place to the right. If it was a zero in front, then we would have removed the zero and pulled everything else one place to the left.

Using continued fractions, we can represent any rational number as a finite-length continued fraction. Irrational numbers, which can't be written in finite form as either regular fractions or decimals, also can't be written in finite form as continued fractions. But it's no worse than (and, in fact, a bit better than) what we do with decimals. In decimals, we write approximations of irrational numbers as a prefix—that is, we write the first digits of the number and stop when it's accurate enough. If we need more accuracy, we add more digits to get a better approximation. With continued fractions, we do the same basic thing: we write an approximation of them using a series of convergents, and if we need more accuracy, we add more convergents.

In fact, it gets a bit better. The continued-fraction form of an irrational number is an infinite series of correction fractions, which we can understand as an infinite sequence of ever-improving approximations of the value of the irrational number. What makes it different from decimals is that with continued fractions, it's easy to generate exactly the next convergent. In fact, we can take the procedure that we used to create a continued fraction and convert it into a function called a *recurrence relation*, which takes the sequence of convergents that we've determined so far and computes the next convergent. That recurrence relation is a beautiful thing about continued fractions: with decimals, there's no way to define a function that produces exactly the next digit.

Another beautiful thing about continued fractions is their precision and compactness. Adding a single convergent to a continued-fraction approximation adds more information than adding a single digit. For example, we know that $\pi = [3; 7, 15, 1, 292, 1, \ldots]$. If we work that out, the first six places of the continued fraction for π would be represented in decimal form as 3.14159265392. That's π correct to the first eleven decimal places. The continued fraction requires five convergents containing a total of eight numerals to get the same precision as eleven digits in decimal form. In general, the conciseness savings of continued fractions is even better than that!

Aside from just being generally cool, continued fractions have some really interesting properties. A very big one is that a lot of numbers become cleaner and clearer in continued fractions. In particular, numbers with no apparent structure or pattern can reveal deep patterns when rendered as continued fractions.

For example, the square root of every nonperfect square integer is an irrational number. They generally don't have any visible structure. For example, the square root of 2 in decimal form is approximately 1.4142135623730951. But if you do it as a continued fraction, you get [1; 2, 2, 2, 2, 2, ...]. All of the square roots of integers that are nonperfect squares have repeated forms in continued fractions.

Another great example is *e*. If you render *e* as a continued fraction, you get $e = [2; 1, 2, 1, 1, 4, 1, 1, 6, 1, 1, 8, 1, 1, 10, 1, 1, 12, 1, \ldots]$. In this and many other cases, continued fractions reveal the underlying structure of the numbers.

There's another cool property of continued fractions. When we write numbers, we write them in a *base*, using powers of a particular number. With decimal numbers, everything is written in powers of 10. For example, 32.12 is $3\times10^1 + 2\times10^0 + 1\times10^{-1} + 2\times10^{-2}$. If we change the number base, we completely change the representation of the number: 12.5 in base 10 becomes 14.4 in base 8. But with continued fractions, the sequence of convergents is exactly the same in different bases!

Doing Arithmetic

Of course, a natural question at this point is, can you actually do arithmetic with these things? They're pretty, and they're interesting, but can you actually use them? Can you do arithmetic with them?

And the answer is heck yes (so long as you're a computer).

For a long time, no one realized that. It took until 1972, when an interesting guy named Bill Gosper came up with a solution.[1] The full details of Gosper's methods are pretty hairy, but the basic idea of the method isn't that hard.

Gosper's fundamental insight was that you could use what we now call *lazy evaluation* to do continued-fraction arithmetic. With lazy evaluation, you don't need to compute the digits of a continued fraction all at once; you compute them one at a time, as you need them.

In modern software terms, you can think of it as being a way of treating a continued fraction as an object with two methods.

1. You can read Gosper's original paper describing his algorithm at http://www.tweedledum.com/rwg/cfup.htm.

In Scala (my personal language of choice), here's how that looks:

cfrac/cfrac.scala

```
trait ContinuedFraction {
  def getIntPart: Int
  def getConvergent: Int
  def getNext: ContinuedFraction

  override def toString: String =
    "[" + getIntPart + "; " + render(1000) + "]"

  def render(invertedEpsilon: Int): String = {
    if (getConvergent > invertedEpsilon) {
      "0"
    } else {
      getConvergent + ", " + getNext.render(invertedEpsilon)
    }
  }

}
```

Using the Scala trait we've just defined, we can implement a continued-fraction object that creates a continued fraction from a floating-point value with this:

cfrac/cfrac.scala

```
class FloatCfrac(f: Double) extends ContinuedFraction {
  def getIntPart: Int =
    if (f > 1) f.floor.toInt
    else 0

  private def getFracPart: Double = f - f.floor

  override def getConvergent: Int =  (1.0/getFracPart).toInt

  override def getNext: ContinuedFraction = {
    if (getFracPart == 0)
      CFracZero
    else {
      val d = (1.0/getFracPart)
      new FloatCfrac(d - d.floor)
    }
  }
}

object CFracZero extends ContinuedFraction {
  def getIntPart: Int = 0
  def getConvergent: Int = 0
  def getNext: ContinuedFraction = CFracZero
  override def render(i: Int): String = "0"
}
```

That code implements exactly the steps of the algorithm we used to compute a continued fraction.

Gosper's second insight was that to get the next convergent of the result of a continued fraction, you only need a finite part of the two continued fractions you are operating on. So you just pull convergents from the fractions that you're working with until you have enough to compute the next convergent of the result.

The actual algorithm is pretty messy. But the gist is this: you can always determine that the next convergent of the result will be within some range, even if that range is as wide as the range from zero to infinity. Each time you get the next convergents of the two operands, you can narrow that range. Eventually, after pulling enough convergents, you'll narrow it to be an integer, which is your convergent. Once you know that convergent, you can then represent the remainder as a new continued fraction in terms of the unknown remainder as an arithmetic operation of the unretrieved convergents of the two operands.

With Gosper's insight, continued fractions become a way of writing numbers that is incredibly well suited to computer programs. Gosper's algorithms for continued-fraction arithmetic give a lot of precision! And they are not biased toward base 10! And it's easy to do computations in a way that allows you to decide on the fly just how much precision you want! It's painful to understand how to implement it, but once you do, the implementation itself is pretty simple, and once it's been implemented, you just use it and it works.

Once you have Gosper's method, continued fractions become beautiful on all counts. Not only do you get all of the cool properties that we've already seen, but arithmetic with continued fractions becomes perfectly precise for as much precision as you want! It's beautiful.

Part IV

Logic

There's a lot more to math than just numbers. The fun of math really starts to become clear when you move beyond arithmetic and get to the abstract stuff. All of the abstractions can be built with two basic things: logic and set theory. So here we'll take a look at logic.

In this part of the book we'll explore what logic is, what a proof is, and what it really means for one statement to logically *follow from* the statements before it. We'll take a look at a couple of logics to see how various logics can describe different kinds of reasoning, and we'll explore the power of logical reasoning by playing with a programming language that's entirely built on logic.

12

Mr. Spock Is Not Logical

I'm a big science fiction fan. In fact, my whole family is pretty much a gaggle of sci-fi geeks. When I was growing up, every Saturday at 6 p.m. was *Star Trek* time, when a local channel showed reruns of the original series. When Saturday came around, we always made sure we were home by 6, and we'd all gather in front of the TV to watch the show. It's a really fond memory. But there's one thing about *Star Trek* for which I'll never forgive Gene Roddenberry or the show: the way that they abused the word "logic," every time Mr. Spock said, "But that would not be logical."

Mr. Spock's pronouncements taught a huge number of people that "logical" means the same thing as either "reasonable" or "correct." When you hear people say that something is logical, what they usually mean *isn't* that it's logical; they mean almost the exact opposite of that—that common sense tells them that it's correct.

If you're using the word "logical" correctly, then just saying that something is logical doesn't mean anything about whether it's correct or not. Anything, *anything at all* can be logical. For something to be logical, it just has to be a valid example of reasoning from some set of premises.

For example, what if I told you this was logical: "If my car is made out of uranium, then the moon is made of cheese." You'd probably think I'm crazy. It's a silly, ridiculous statement, and by intuition, it's not true. Mr. Spock would certainly say that it's not logical. But, in fact, it *is* logical. Under the standard rules of predicate logic, it's a true statement.

The fact that it's a stupid statement that doesn't mean anything is irrelevant. It's logical because logic says it's true.

How is it logical? To say that something is logical, what you're saying is that it is something that can be inferred or proven using a formal system of reasoning. In *first-order predicate logic* (FOPL)—the most common logic used by mathematicians—an if/then statement (formally called an *implication*) is true if either the "if" part is false or the "then" part is true. If you put something false in the "if" part, you can put *anything at all* in the "then" statement, and the if/then statement will be *logically true*.

But there's more to it than that. Logic, in the sense that we generally talk about it, isn't really one thing. Logic is a name for the general family of formal proof systems with inference rules. There are many logics, and a statement that is a valid inference (that is, is logical) in one system may not be valid in another. To give you a very simple example, most people know that in logic there's a rule called the *law of the excluded middle*, which says for a given statement *A*, that *A or not A* is true. In FOPL, that's a kind of statement called a *tautology* that must always be true. But there are more kinds of logic that we can use. There's another very useful logic called *intuitionistic logic*, in which *A or not A* is *not* necessarily true. You cannot infer anything about whether it's true or false without proving whether *A* is true or false.

Logic has a lot more variety than you might expect. FOPL and intuitionistic logic are pretty similar, but there are *many* other kinds of logic that are useful for different things. FOPL is great for proofs in algebra and geometry, but it's *awful* at talking about time. FOPL has no good way to say something like "I won't be hungry until 6 p.m. tonight" that really captures the temporal meaning of that statement. But there are logics like CTL (which we'll talk more about in 15, *Temporal Reasoning*, on page 117) that are designed specifically to be good at that kind of statement. But they're not useful for the kinds of things that FOPL is good at. Each logic is designed for a purpose, to do a particular kind of reasoning. Each logic can be used in different ways to prove different things.

What *Star Trek* and Mr. Spock get right is that one of the fundamental points of logic is being unbiased. Mr. Spock does, at least in theory, try to reduce things to a form where they can be understood without emotion. But saying that you're being logical isn't enough to be meaningful. You need to say *which logic* you're using. You need to say what axioms (basic facts) you're using to reason with and how your logic allows you to do your reasoning.

The same argument can be logically valid and correct in one logic and logically invalid and incorrect in another logic, or even in the same logic with a different set of basic facts. Without specifying the logic and the axioms, saying that a conclusion is logical tells you nothing.

What Is Logic, Really?

A logic is a system for mechanical reasoning. It's a system that allows you to express an argument in a neutral symbolic form and then use that symbolic form to determine whether or not the argument is valid. What matters in logic isn't meaning; instead it's whether or not the steps of reasoning that make up the argument follow from one another. It's a very powerful tool, specifically because it ignores meaning when it's doing reasoning. Logic can't be biased toward one side of an argument, because logic neither knows nor cares what the argument means!

To make that kind of unbiased reasoning work, a logic needs to be structured in a rigorous, formal way. That formality shows us that the reasoning (formally called *inference*) in the logic will produce correct results when applied to real arguments.

We care about this because logic is at the heart of how we reason, and reasoning is at the heart of how we communicate! Every political debate, every philosophical argument, every technical paper, every mathematical proof, every *argument* at its core has a logical structure. We use logic to capture that core in a way that lets us see the structure, and more than just see it—understand it and test it and ultimately see if it's correct.

We make it all formal because we want to be sure that when we're analyzing an argument, we aren't being biased. The formality makes sure that our logical system is really reliable, systematic, and objective. What logic does is take an argument and reduce it to a symbolic form that can be evaluated for correctness *without knowing what it means*. If you consider meaning, it's easy to let your intuitive understanding sway you. If you know what the argument is about, it's easy to let your own biases creep in. But logic reduces it to symbols that can, in theory, be evaluated by a machine.

Formally, a logic consists of three parts: syntax, semantics, and a set of inference rules. The syntax tells you exactly what statements in the logic look like and how you should read and write them. Semantics tells what statements in the logic mean by showing you how to move between the abstract symbolic statements of the logic and the ideas and objects that you're reasoning about. Finally, the inference rules describe how, given a collection of statements written in the logic, you can perform reasoning in the logic.

FOPL, Logically

The first thing you need to know about a logic is its *syntax*. A logic's syntax tells you, formally, how to read and write sentences in that logic. Syntax doesn't tell you how to *understand* the statements or what they mean. It just tells you what they look like and how to correctly put them together.

Logic separates the meaning of the statements from the syntax. The meaning is called the *semantics* of the logic. For FOPL, the semantics are pretty easy to follow—not because they're really all that simple, but because you use them every day. Most arguments, most bits of reasoning that you hear every day, are FOPL, so you're used to it even if you don't realize it.

We're going to walk through the basic syntax and semantics of FOPL together because it's much easier to follow that way. We'll start with the objects that we're going to reason about.

The point of logic is to let you reason about things. Those things can be concrete things like cars or people, or abstract entities like triangles or sets. But to reason in a logic, we need to use symbols to represent those objects. For each object that we want to reason about we introduce a symbol called a *constant*, or an *atom*. Each atom represents a particular object, number, or value that can be reasoned about using the logic. If I wanted to reason about my family, the constants will be the names of members of my family, the places we live, and so on. I'll write constants as either numbers or quoted words.

When you don't want to refer to a specific object, you can use a *variable*. For example, if you want to say that every object has a property, like every person has a father, you don't want to have to write "Mark has a father," "Jennifer has a father," "Rebecca has a father," and so on. You want to be able to write one statement that says that it's true for everyone. Variables let you do that. A variable doesn't have any meaning on its own; it gets its meaning from context. We'll see what that means in a little while, when we get to *quantifiers*.

The last type of thing that we need for FOPL is a *predicate*. A predicate is sort of like a function that describes properties of objects or relationships between objects. For example, when I say "Mark has a father," the phrase "has a father" is the predicate. In our examples, we'll write predicates using an uppercase identifier, with the objects it's talking about following inside parens. If I wanted to write the statement "Mark's father is Irving" in FOPL, I would write *Father("Irving", "Mark")*.

Every predicate, when it's defined, explains when it's true and when it's false. If I wanted to use my example predicate, *Father*, my definition of it would need to explain when it's true. In this case, you might think that the name is enough: we know what it means when we say "Joe is Jane's father." But logic needs to be precise. If Joe is actually Jane's adoptive father, should *Father("Joe", "Jane")* be true? If what we care about is family relationships, then the answer should be yes; if what we care about is biological relationships, then the answer should be no. For our examples, we'll say that we're talking about biological relationships.

A predicate with its parameters filled in is called either a *simple fact* or a *simple statement*.

We can form more interesting statements by taking the statements that we have and either modifying them or combining them. The simplest combinations and modifications are "and" (called *conjunction*), "or"(called *disjunction*), and "not" (called *negation*). In formal syntax, we use the $\land$ symbol for "and," the $\lor$ symbol for "or," and the $\neg$ symbol for "not." Using these, we can write things like *Father("Mark", "Rebecca") ∧ Mother("Jennifer", "Rebecca")* (Mark is Rebecca's father and Jennifer is Rebecca's mother), *YoungestChild("Aaron", "Mark") ∨ YoungestChild("Rebecca", "Mark")* (Aaron is Mark's youngest child or Rebecca is Mark's youngest child, and *¬Mother("Mark", "Rebecca")* (Mark is not Rebecca's mother).

And, *or*, and *not* all work *almost* the way you expect them to. $A \land B$ is true if both A is true and B is true. $A \lor B$ is true if A is true or if B is true. $\neg A$ is true if A is false.

There's one tricky thing about the logical *or*. In informal talk, if you say "I want the hamburger or the chicken sandwich," what you mean is that you want *either* the hamburger *or* the chicken sandwich, but not both. With a logical *or*, $A \lor B$, is true whenever *at least* one of A and B is true. It's also OK for *both* to be true. The informal sense of *or* is called an *exclusive or* in FOPL. When we're defining the logic, we don't define the exclusive *or* because it can be written in terms of the other statements.[1]

With $\land$, $\lor$, and $\neg$, we can form all of the sentences that we want to. But we're missing one thing that we really want: *if-then*, also known as *implication*. In arguments of all types, implication is a common and useful tool, so we'd really like to be able to write it directly. Strictly speaking, we don't need it because we can write it using *and*, *or*, and *not*. But because it's so useful, we'll add it anyway. There are two kinds of implications: *simple if* and *if-and-only-if*.

A *simple if* statement is written $A \Rightarrow B$, which can be read as either "A implies B," or "if A then B." What it means is that

---

1. The statement *A exclusive or B* can be written $A \lor B \land \neg(A \land B)$.

if the A part is true, then the B part must also be true—and also the inverse: if the B part is false, then the A part must also be false.

If-and-only-if is written $A \Leftrightarrow B$, which is read "A if-and-only-if B," or "A if/f B" for short. *If/f* is the logical version of equality: $A \Leftrightarrow B$ is true whenever A and B are both true or whenever A and B are both false. As the double-headed arrow notation suggests, $A \Leftrightarrow B$ is exactly the same as $A \Rightarrow B \wedge B \Rightarrow A$.

The connectives give us the ability to form all of the basic logical statements. But we still can't write interesting arguments. Simple statements like these just aren't enough. To see why, let's look at an example of one of the simplest and most well-known logical arguments that came to us from the ancient Greek philosopher Aristotle:

1. All men are mortal
2. Socrates is a man.
3. Therefore, Socrates is mortal.

With what we know about FOPL so far, we can't write that argument. We know how to write specific statements about specific atoms, which means that we can write steps 2 and 3 as *Is_A_Man("Socrates")*, and *Is_Mortal("Socrates")*. But we can't write that first statement. We have no way of saying "All men are mortal" because we have no way of saying "all men." That's a statement about *all* atoms, and we have no way yet of making a general statement about all atoms.

To make a general statement about all possible values, you use a *universal statement*. It's written $\forall a: P(a)$ (read "for all a, $P(a)$"), which means that $P(a)$ is true for all possible values of a.

Most of the time, universal statements show up in implications, which let you limit the statement to a particular set of values instead of allowing all possible values. In the mortal Socrates example, the universal statement "All men are mortal" would be written as $\forall x: Is\_A\_Man(x) \Rightarrow Is\_Mortal(x)$ ("For all x, if x is a man, then x is mortal"). Since this is used so frequently, there's a shorthand: we'll often write it as $\forall x \in Is\_A\_Man: Is\_Mortal(x)$.

The last kind of thing we want to be able to do is called an *existential* statement. An existential lets us say that there must be a value that makes a statement true, even if we don't know exactly what that statement is. Using our family statements, we can say that I must have a father: $\exists x$: *Father(x, "Mark")*, which we read as "There exists an x such that x is the father of Mark."

Existentials are most useful when they're combined with universals. Saying that I must have a father isn't a particularly useful statement: we know who my father is. But by combining a universal with an existential, I can say that *everyone* has a father: $\forall x: \exists y: Father(y, x)$. That's always true—it's a fundamental fact of human biology. If we see a person, we know that that person must have a father. We may not know who he is. In fact, the person may not know who his or her father is. But we know, without question, that this person *has* a father. And that combination of quantifiers lets us say that.

Show Me Something New!

Now we've seen the language of logic: how to read and write statements and how to understand what they mean. But all that we have so far is the language for writing statements. What makes it into a logic is the ability to *prove* things. Proof in logic is done by *inference*: inference gives you a way of taking what you know and using it to prove new facts, adding to what you know.

I'm not going to go through the entire set of inference rules allowed in FOPL in detail in this section. I'm just going to give you a couple of examples that are enough to demonstrate a bit of inference. In the next chapter, I'll show you all of the rules in a more detailed form that's useful for checking proofs. But for now, here's a few to get the sense of how it works:

Modus Ponens This is the most fundamental rule of predicate logic. If I know that $P(x) \Rightarrow Q(x)$ ($P(x)$ implies $Q(x)$) and I know $P(x)$, then I can infer that $Q(x)$ must be true.

Likewise, we can do the reverse. If I know that $P(x) \Rightarrow Q(x)$, and I know that $\neg Q(x)$ (that $Q(x)$ is false), then I can conclude that $\neg P(x)$; that is, that $P(x)$ must be false.

Weakening If I know that $P(x) \land Q(x)$ is true, then I can infer that $P(x)$ must be true.

Similarly, if I know $P(x) \lor Q(x)$ is true and I know that $Q(x)$ is false, then $P(x)$ must be true.

Universal Elimination If we know that $\forall x: P(x)$ is true and *"a"* is a specific atom, then we can infer that *P("a")* is true.

Existential Introduction If x is an unused variable and we know that *P("a")* is true, then we can infer that $\exists x: P(x)$ is true.

Universal Introduction If we can make a proof that shows that $P(x)$ is true without knowing anything about x, then we can generalize from that and say that $\forall x: P(x)$.

To reason with a logic, you start with a set of axioms, which are the basic facts that you know are true even though you don't have a proof. A statement is true in the logic if that statement can be proven by using axioms and the inference rules of the logic.

So once again, here's a set of axioms about my family.

- Axiom 1. *Father("Mark", "Rebecca")* Mark is Rebecca's father.
- Axiom 2. *Mother("Jennifer", "Rebecca")*
- Axiom 3. *Father("Irving", "Mark")*
- Axiom 4. *Mother("Gail", "Mark")*
- Axiom 5. *Father("Robert", "Irving")*
- Axiom 6. *Mother("Anna", "Irving")*
- Axiom 7. $\forall a, \forall b:(Father(a, b) \lor Mother(a, b)) \Rightarrow Parent(a, b)$
- Axiom 8. $\forall g, \forall c : (\exists p : Parent(g, p) \land Parent(p, c)) \Rightarrow Grandparent(g, c)$

Now let's use these axioms and our inference rules to prove that Irving is Rebecca's grandparent.

Example: Prove that Irving is Rebecca's grandparent.

1. Since we know by axiom 1 that *Father("Mark", "Rebecca")*, we can infer *Parent("Mark", "Rebecca")*. We'll call this inference I1.

2. Since we know by axiom 3 that *Father("Irving","Mark")*, we can infer *Parent("Irving","Mark")*. We'll call this inference I2.

3. Since we know by I1 and I2 that *Parent(Irving, Mark)* and *Parent("Mark", "Rebecca")*, we can infer *Parent("Irving", "Mark")* ∧ *Parent(Mark,Rebecca)*. We'll call this inference I3.

4. Since by I3, we know *Parent("Irving", "Mark")* ∧ *Parent("Mark", "Rebecca")* using axiom 8, we can infer *Grandparent("Irving", "Rebecca")*.

5. QED.

In a given logic, a chain of inferences forged from its rules is called a *proof*. The chain in our example is a proof in first-order predicate logic. A very important thing to notice is that the proof is entirely symbolic: we don't need to know what the atoms represent or what the predicates mean! The inference process in logic is purely symbolic and can be done with absolutely no clue at all about what the statements that you're proving mean. Inference is a simple process that uses inference rules to work from a given set of premises to a conclusion. Given the right set of premises, you can prove *almost any* statement; given a choice of both premises *and* logic, you can prove *absolutely any* statement.

Let's try to redo the inductive proof from 1, *Natural Numbers*, on page 3, this time being clear about how we're using logical inference.

Example: Prove that for all n, the sum of the natural numbers from 0 to n is n(n+1)/2.

1. The induction rule, in logical terms, is an implication that says this: *if* that statement is both true for 0 *and* if we can show that for all values n greater than or equal to 1, it's true for n if it was true for $n - 1$, then it will be true for all n—and we'll have proven it.

2. We need to show those two cases in order to use the implication. We start with the base case. We need to show that $n(n+1)/2$ = sum of the naturals from 0 to n when $n = 0$. If we wanted to be complete, we would actually need to reason through the formal Peano-based definitions of addition and multiplication. In the definition of multiplication, it says that zero times anything is zero; so we can use that as a logical equivalence and substitute 0 for $n(n+1)$—so the sum from 0 to 0 is 0, and the base case is proven. We now have as an inferred fact that it's true for 0.

3. Now comes the inductive part. The inductive part is itself an implication. If we take the fact that we're trying to prove as a predicate P, then what we want to do is show that for all n, $P(n)$ implies $P(n+1)$. We do that by using the universal introduction inference rule. We show that the $n/n+1$ inference is true by working through it with an *unbound n*:

   Suppose, for a number n, that it's true. Now we want to prove it for $n + 1$.

   So what we want to prove is this:

$$(0 + 1 + 2 + 3 + \cdots + n + n + 1) = \frac{(n+1)(n+2)}{2}$$

4. Then we go through the algebraic part exactly as we did the first time through. At the end of that, we have a statement that if P is true for n, then it's true for $n + 1$. We can use an inference rule to generalize that into a universal.

5. Now we have both of the statements that were required by the implication, and as a result, we can use implication inference (the rule that if A implies B and we know A is true, then B must be true). We know that P is true for 0. We know that for all $n > 0$, P is true for n if it was true for $n - 1$. So by the induction rule, we can now infer the conclusion: For all natural numbers n, the sum of the numbers from 0 to n is $n(n+1)/2$.

6. QED.

Going through all of the inference rules here might seem like it's a bit of overkill, but there is a point to it.

The beautiful thing about logic is that it's a simple way of making reasoning clear and mechanical. The list of rules can seem complex, but when you think about it, virtually every argument that you hear in daily life, in politics, in business, in school, or around the dinner table, when you boil it down, are all expressible in first-order predicate logic. And all of those arguments, every one, in every case, are *testable* by turning them into the FOPL form and then using that set of inference rules. That's all you need to be able to check any argument or make any proof. When you think of it that way, you should be able to start to see why it really is amazingly simple.

13

Proofs, Truth, and Trees: Oh My!

The point of logic is to make it possible to prove things. What is a proof? In math, it's a sequence of logical inferences that show how a new fact, called a *conclusion*, can be derived from a set of known facts, called *premises*.

Proofs tend to scare people. I can still remember being a sophomore in high school and being introduced to proofs in my geometry class. It was by far the worst experience I'd ever had in a math class! I'd try to work out a proof in my homework; and when the teacher handed it back, it would be covered in red ink, with every other line marked "doesn't follow," or "missing case." I just couldn't figure out what I was supposed to be doing.

A lot of people have had pretty much the same experience. We're taught that a valid proof has to have every step follow from what came before, and it has to cover every possible case. Sadly, what we aren't taught nearly as well is exactly how to tell when a step *follows*, or how to be sure that we've covered all of the cases. Understanding how to do that only comes from actually understanding the logic and the mechanics of logical inference used in a proof.

That's what makes proofs so hard for so many of us. It's not really that proofs are so terribly hard. Instead, it's the fact that mastering proofs requires mastering the logic that's used in those proofs. The *logic* part of proofs isn't really taught to us in most math classes. It's treated as something that we should just understand. For some people, that's easy; for many of us, it isn't.

So in this chapter, we're going to look at the mechanics of proofs in FOPL. I'm going to approach it using a technique called *truth trees* or *semantic tableaus*. What you'll see is a bit different from the typical presentation of truth trees. I was lucky enough to take a course in logic when I was in college from a great professor named Ernest Lepore. Professor Lepore had worked out his own way of teaching truth trees to students, and to this day I still use his. Since they're what I use, they're what I'm going to show you. (If you want to learn more about logic and proofs from a less mathematical perspective, then you can't go wrong with Professor Lepore's book, *Meaning and Argument: An Introduction to Logic Through Language [Lep00]*.

Building a Simple Proof with a Tree

Truth trees work by creating a contradiction. We take a statement that we want to prove true, turn it around by negating it, and then show that that negation leads to contradictions.

For example, I can use this basic method to prove that there isn't a largest even number, N:

Example: Prove there is no such thing as a largest even number, N.

1. We'll start by inverting the fact to prove, and then we'll show that it leads to a contradiction. Suppose that there *is* a largest even number, N.
2. Since N is the largest even number, then for every other even number n, $n < N$.
3. N is a natural number; therefore, it can be added with other natural numbers.
4. If we add 2 to N, the result will be even and it will be larger than N.
5. We just created an even number larger than N, which contradicts step 2. Since we have a contradiction, that means that the statement that N is the largest even number is false.

That's how truth trees work: by showing that every possible chain of reasoning starting from a negation of what you want to prove will lead to a contradiction. The tree mechanism gives you an easy, visual way of making sure that you've got all of your cases covered: every time you do anything that introduces another case, you create a branch of the tree.

In practice, I don't actually use truth trees for writing proofs; I use them for *checking* a proof. When I think that I've got a good proof of something in FOPL, I check that I got it right by throwing together a truth tree to check that I didn't make any mistakes. The tree makes it easy to look at a proof and verify that everything follows correct inference and that all cases are covered, ensuring that the proof is valid.

The way that you do that in a truth tree is to start by writing down the premises that you're given in a column; then you take the statement that you want to prove and write it at the top of the page. Then you *negate* that statement. In the truth tree we'll show that with the statement negated, every possible inference path leads to a contradiction. If that happens, then we know that the negated statement is false, and the statement that we wanted to prove is true.

Implication Equivalence	$A \Rightarrow B$ is equivalent to $\neg A \lor B$
Universal Negation	$\neg \forall x: P(x)$ is equivalent to $\exists x: \neg P(x)$
Existential Negation	$\neg \exists x: P(x)$ is equivalent to $\forall x: \neg P(x)$
And Negation	$\neg(A \land B)$ is equivalent to $\neg A \lor \neg B$
Or Negation	$\neg (A \lor B)$ is equivalent to $\neg A \land \neg B$
Double Negation	$\neg \neg A$ is equivalent to A
Universal Reordering	$\forall a: (\forall b: P(a, b))$ is equivalent to $\forall b: (\forall a: P(a, b))$
Existential Reordering	$\exists a: (\exists b: P(a, b))$ is equivalent to $\exists b: (\exists a: P(a, b))$

Table 1—Logical equivalence rules

Given	Rule	Infer
And Weakening (left)	$A \land B$	A
And Introduction	A and B	$A \land B$
Or Branching	$A \lor B$	Two branches: one with A, one with B
Or Negation	$A \lor B$ and $\neg A$	B
Or Introduction	A	$A \lor B$
Modus Ponens	$A \Rightarrow B$ and A	B
Universal Elimination	$\forall x: P(x)$	$P(a)$ for any specific atom a
Existential Elimination	$\exists x: P(x)$	$P(a)$ for any unused atom a

Table 2—The inference rules for truth trees in FOPL

Instead of droning on through a list of the inference rules, I've just listed them in tables. For the most part, if you look at them carefully and think about them, you can figure out what they mean. But the point of logic as a reasoning system is that you *don't need to know what they mean*. The logic defines these inference rules; and as long as you follow them, you'll be able to create valid proofs.

A Proof from Nothing

Now that we've seen all of the sequents, how can we use them?

Let's start with a proof of a simple but fundamental rule of logic called the *law of the excluded middle*. It says that any statement must be either true or false. When you write that in logic, it means that if you have a statement A, then regardless of what A is, $A \lor \neg A$ must be true.

The law of the excluded middle is a *tautology*, which means that it's a fundamental truth that must always be true, regardless of what axioms we choose. In order to show that the tautology is true, we need to build a proof that uses nothing but the statement, the logic's rules of inference. If we can derive a proof of $A \lor \neg A$ with no premises, we can show that it's a universal truth.

Since we want to prove $A \vee \neg A$, we'll start the truth tree with its negation, and then we'll show that every path through the tree will end with a contradiction. The proof is shown in the following figure.

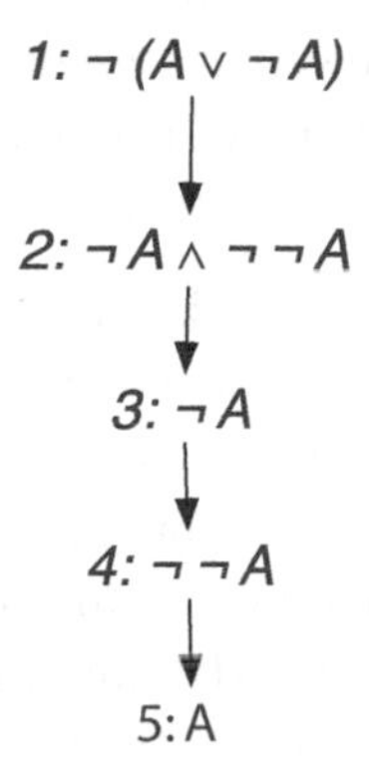

Figure 10—A proof tree for the law of the excluded middle

Prove that $A \vee \neg A$.

1. We want to prove $A \vee \neg A$, so we start with its negation: $\neg(A \vee \neg A)$.
2. By using the *and negation* equivalence rule, we can write down $\neg A \wedge \neg\neg A$.
3. By using *and weakening* on (2), we can add $\neg A$.
4. By using *and weakening* on (2) again, we can add $\neg\neg A$.
5. By using *double negation* on (4), we can add A to the tree.
6. Our only branch of the tree now contains both A and $\neg A$, which is a contradiction.

That's a complete proof of our tautology. It wasn't so hard, was it? The point of the proof is that we know that we've covered all of our bases, everything follows from what came before, and all cases are definitely covered.

Notice that in each step, we don't need to think about what anything means. We can just observe that at the beginning of the step we have something that matches a pattern in one of the rules and we use that rule to derive something new: a new *inference,* which added to the tree.

This was an easy example, but even in a case like this, the difficult part of a proof isn't applying the rules. Applying an inference rule is easy. But how do you decide which inference rule to apply? That's the hard part. How did I know to use *and weakening* in step 3? Basically, I made an educated guess. I knew that I wanted to get to a simple contradiction, and to do that, I needed to separate the pieces that could turn into a contradiction.

If you're a programmer, building a proof is a lot like walking through a search tree. In theory, at each step in the proof, you can find every inference rule that's applicable and try using each one. If you keep doing that and the statement is provable, then eventually you'll find the proof. It may take a very long time, but if the statement is provable, you'll get there. (But as we'll see in 27, *The Halting Problem*, on page 253, if the statement isn't provable, you might be searching forever.) In practice, building real proofs is a combination of a search process and a lot of directed trial and error. You look at what you want to prove and what facts you have available to you, and you figure out what possible inference steps you can take. Knowing what your options are, you pick the one that seems, based on your understanding of what you're proving, to be the most likely to lead to your conclusion. If you try that and it doesn't work, then you pick a different option and try again.

All in the Family

Now let's move on and try something more interesting. We're going to go back to the family-relationships example from the previous chapter and prove that if two people are cousins, then they have a common grandparent.

In our family relations, we need to define what cousins are in logical terms: we'll say that two people are cousins if they each have a parent who's a sibling of one of the other's parents. In FOPL, that's this: $\forall a: \forall b: Cousin(a, b) \Leftrightarrow \exists m: \exists n: Sibling(m, n) \land Parent(m, a) \land Parent(n, b)$.

We want to prove that $\forall d: \forall e: Cousin(d, e) \Leftrightarrow \exists g: Grandparent(g, d) \land Grandparent(g, e)$.

Like the proof of the excluded middle, we can do this without branching the tree. The proof does have a fair bit of jumping around. The key to understanding it is to remember that anything *above* a statement in the tree is fair game for use below it.

Throughout this proof, our goal is to take the definitions of *Cousin* and *Grandparent*, decompose them into simple terms, and then push those terms to form a contradiction.

1. As always, we start by taking the statement we want to prove and negating it; that's the root of the tree:

   $\neg (\forall d: \forall e: Cousin(d, e) \Leftrightarrow \exists g: Grandparent(g, d) \land Grandparent(g, e).$

2. Use *universal negation equivalence* to push the $\neg$ inside the first $\forall$:

   $(\exists d: \neg \forall e: Cousin(d, e) \Leftrightarrow \exists g: Grandparent(g, d) \land Grandparent(g, e)).$

3. Use *universal negation equivalence* again to push inside the second $\forall$:

   $\exists d: \exists e: \neg (Cousin\ (d, e) \Leftrightarrow \exists g(Grandparent(g, d) \land Grandparent(g, e)).$

4. Use *implication equivalence* to convert $\Rightarrow$ to an $\lor$:

   $\exists d: \exists e: \neg (\neg Cousin(d, e) \lor \exists g: Grandparent(g, d) \land Grandparent(g, e)).$

5. Use *or negation* to push the $\neg$ into the *or*:

   $\exists d, \exists e: Cousin(d, e) \land \neg \exists g: Grandparent(g, d) \land Grandparent(g, e)).$

6. Use *existential elimination* to get rid of the $\exists$ by substituting a new variable:

   $Cousin(d', e') \land \neg \exists g: Grandparent(g, d') \land Grandparent(g, e')).$

7. Use *and weakening* to separate the left clause of the $\land$:

   $Cousin(d', e').$

8. Use *universal elimination* to specialize the definition of cousin with d' and e':

   $Cousin(d', e') \Leftrightarrow \exists p, \exists q: Sibling(p, q) \land Parent(p, d') \land Parent(q, e').$

9. *Modus ponens*:

   $\exists p, \exists q: Sibling(p, q) \land Parent(p, d') \land Parent(q, e').$

10. *Existential elimination*:

    $Sibling(p', q') \land Parent(p', d') \land Parent(q', e').$

11. *And weakening*:

    $Sibling(p', q')$.

12. *Universal elimination* to specialize the definition of sibling for p', q':

    $Sibling(p', q') \Rightarrow \exists g: Parent(g, p') \land Parent(g, q')$.

13. *Modus ponens*:

    $\exists g: Parent(g, p') \land Parent(g, q')$.

14. *Universal elimination* to specialize the definition of *Grandparent* for d':

    $\forall g: Grandparent(g, d') \Leftrightarrow \exists e, Parent(g, e) \land Parent(e, d')$.

15. *And introduction*:

    $Parent(p', d') \land Parent(g, p')$.

16. *Modus ponens*:

    $Grandparent(g, d')$.

17. Now repeat the specialization of *Grandparent* using e', and you'll get

    $Grandparent(g, e')$.

18. Go back to where we did *and weakening* to separate the left-hand clause and do a *separate right* to separate the right, and you'll get

    $\neg \exists g: Grandparent(g, d') \land Grandparent(g, e'))$.

19. And that is a contradiction. Since we never branched, we've got a contradiction on the only branch that our tree has, and that means we're done.

Branching Proofs

The two truth-tree proofs we've seen so far have both been single-branch proofs. In practice a lot of interesting proofs, particularly in number theory and geometry, turn out to be single-branch. Some proofs, though, do need branching. To see how that works, we're going to look at another tautology: the transitivity of implication. If we know that statement A implies statement B, and we also know that statement B implies a third statement, C, then it must be true that A implies C. In logical form, $(A \Rightarrow B \land B \Rightarrow C) \Rightarrow (A \Rightarrow C)$.

The following figure shows the truth-tree proof of this. We'll walk through the steps together. The strategy in this proof is similar to what we did in our proof of the law of the excluded middle. We want to just take the statement at the top of the proof, decompose it into simple statements, and then try to find contradictions.

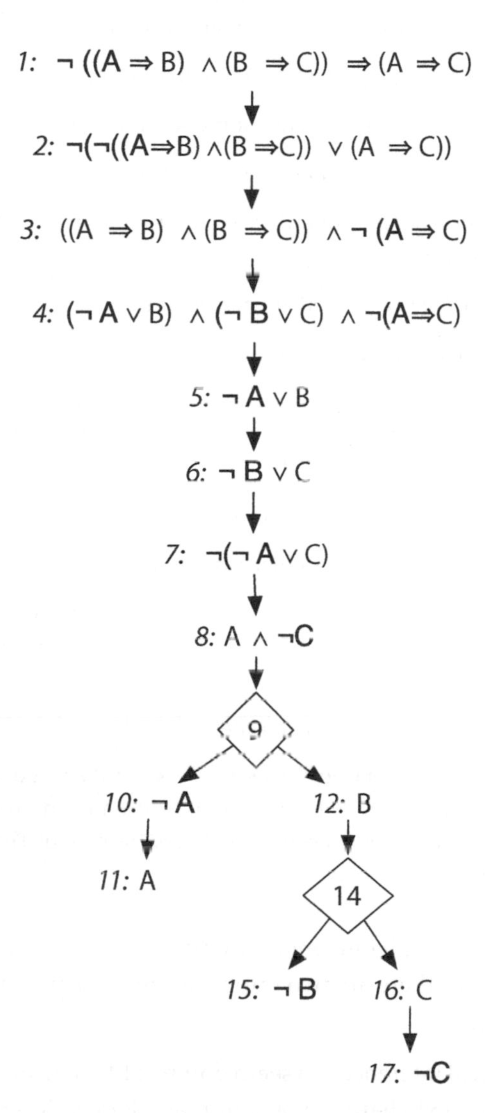

Figure 11—Truth tree example

Example: Prove that $(A \Rightarrow B \wedge B \Rightarrow C) \Rightarrow (A \Rightarrow C)$.

1. As always, start by negating the statement we want to prove.
2. Use *implication equivalence* to get rid of the outermost implication.
3. Use *and negation* and *double negation* to push the outer negation through the statement.
4. *Implication equivalence* twice to eliminate the first two implications.
5. *And weakening* to separate out the first term.
6. *And weakening* to separate out the second term.
7. *And weakening* to separate out the third term.
8. *Implication equivalence* and *or negation* to simplify the statement from step 7.
9. *Or branching* on the statement from step 5.
10. Left branch of the *or branching* from step 9.
11. *And weakening* of the statement in step 8. This gives us both A and $\neg A$ in this branch, which is a contradiction. That means we're done on this branch.
12. Right branch of the *or branching* from step 9.
13. *Or branching* of the statement in step 6.
14. Left branch of the *or branching* in step 13. That gives us $\neg B$ here and B in step 12, which is a contradiction, so we're done on this branch.
15. Right branch of the *or branching* in step 13.
16. *And weakening* of step 8. This gives us $\neg C$. We've got C in step 16, which means that we've got a contradiction on our final branch. All of the branches end in contradiction! That means we're done.

This dive through inference with truth trees has been pretty detailed, but there are two important reasons why we went through it.

First, logical inference can seem downright magical. As we'll see in the next chapter, you can make logical inference do really difficult things. That makes it seem like inference must be doing something complicated. But it isn't. Inference is easy. You can see that there's a small number of rules, and you can understand them all without much effort. But this simple framework turns into something astonishingly powerful.

Second, seeing the rules in detail helps to make one of the fundamental points of logic really clear. These rules are completely syntactic, so you can do the reasoning in terms of symbols, not in terms of meaning.

14

Programming with Logic

Having seen a bit of how inference in first-order predicate logic works using truth trees, you might wonder just how much you can do with inference. The answer is, a whole heck of a lot! In fact, anything that a computer can do—any computation that can be written as a program—can be done by inference in first-order predicate logic!

This isn't just an empty boast or a theoretical equivalence, but a real and practical fact. There's a very powerful, very useful programming language called Prolog, in which a program consists of nothing but collections of facts and predicates—all you can do is provide it with a collection of facts and predicates. The execution of a Prolog program is a sequence of inferences performed by the Prolog interpreter using the facts and predicates of your program.

Odds are that even if you're a professional engineer, you probably won't ever need to write Prolog. But I'm a programming-language junkie, and trust me on this: it's worth learning about. I know a positively silly number of programming languages, and when someone asks me what one language they should learn, I usually tell them Prolog. Not because I expect them to use it either, but Prolog opens your mind to a completely different way of programming, and it's worth the effort just for that.

I'm not going to teach you all that you'd want to know about Prolog—that's far too much to cover here. For that, you need to read a book about Prolog. (I list a couple of texts in the references at the end of this section.) But I can definitely give you a taste to see what it's like.

Computing Family Relationships

I'm going to introduce you to Prolog by using the same example as the previous chapter: family relationships. But instead of writing them out in formal logical syntax, we'll do it in Prolog, and we'll see how you'd interact with the Prolog system. I'm using a free, open-source Prolog interpreter called SWI-Prolog,[1] but you can use that or any other Prolog system that runs on your computer.

In logic, you reason over objects, called *atoms*. In Prolog, an atom is any *lowercase* identifier, number, or quoted string. For example, mark, 23, and "q" are atoms. For an example, we'll look at reasoning about family relationships, so the atoms will all be first names.

A variable in Prolog is a symbol that represents an atom. Variables can be used in the logic to reason about universal properties—if every object has a property (like, for example, every person has a father), there's a way of using a variable to say that in the logic. Variables in Prolog are uppercase letters: X, Y.

A predicate in Prolog is a statement that allows you to define or describe properties of objects and variables. A predicate is written as an identifier, with the objects it's talking about following inside parens. For example, I can say that my father is Irving using a predicate named father: father(Irving, Mark). So in our example, we state a series of facts using a predicate person to say that a bunch of atoms are people. What we're doing here is presenting the Prolog interpreter with a set of basic, known facts, which are called the *axioms* of our logic.

logic/family.pl
```
person(aaron).
person(rebecca).
person(mark).
person(jennifer).
person(irving).
person(gail).
person(yin).
person(paul).
```

1. You can get information and download SWI-Prolog from http://www.swi-prolog.org/.

```
person(deb).
person(jocelyn).
```

We can connect a series of facts using commas: A, B means that both A and B are true.

Now we can check to see if certain atoms are people. If we load the set of facts into Prolog, we can ask it questions:

```
⇒ person(mark).
❮ true.
⇒ person(piratethedog).
❮ false.
⇒ person(jennifer).
❮ true.
```

At this point, all that it can tell us is whether these statements were explicitly given to it as facts. That's good, but it's not particularly interesting. With what we've given it so far, there's nothing that it can infer. In order to give it meat for inference, we can define more interesting predicates. We're going to give it rules that it needs in order to infer family relationships. Before we do that, though, we need to give it some fundamental family relationships. To do that, we need to give it some more complicated facts using multiple parameter predicates. We'll do that by writing facts about fathers and mothers.

logic/family.pl

```
father(mark, aaron).
father(mark, rebecca).
father(irving, mark).
father(irving, deb).
father(paul, jennifer).
mother(jennifer, aaron).
mother(jennifer, rebecca).
mother(gail, mark).
mother(gail, deb).
mother(yin, jennifer).
mother(deb, jocelyn).
```

Now, finally, we can get to the interesting part. What makes logic valuable is inference—that is, the ability to take known facts and combine them in ways that produce new facts.

So suppose that we want to be able to talk about who is a parent and who is a grandparent. We don't need to go through and enumerate all of the parent relationships. We've already said who is whose mother and father. So we can describe being a parent in terms of being a mother or a father. And once we've done that, we can describe being a grandparent in terms of being a parent:

logic/family.pl
```
grandparent(X, Y) :- parent(X, Z), parent(Z, Y).
```

First, we have two lines that define what a parent is in terms of logic. Each line defines one alternative: the two parent lines can be read as "*X* is *Y*'s parent if *X* is *Y*'s father, or *X* is *Y*'s parent if *X* is *Y*'s mother."

Then we define what a grandparent is logically in terms of parents: "*X* is *Y*'s grandparent if there's a *Z*, where *X* is *Z*'s parent, and *Z* is *Y*'s parent." I can ask the Prolog interpreter to infer facts about parents and grandparents. To do that, I just write logical statements with variables, and it will try to find values for those variables for which it can infer that the statement is true.

What if we wanted to write a rule to define what it means to be a sibling? A sibling is someone who has a common parent. So we can write that logically in Prolog by saying that `A` and `B` are siblings if there's a `P` that is the parent of both `A` and `B`:

logic/family.pl
```
sibling1(X, Y) :-
  parent(P, X),
  parent(P, Y).
```

Let's try that rule out.

```
?-
  sibling(X, rebecca)
X = aaron ;
X = rebecca ;
X = aaron ;
X = rebecca ;
false.
```

Prolog will infer any fact that fits the logical rules. The compiler neither knows nor cares whether it makes sense: it's a computer program that does exactly what you tell it. Using common sense, we know that it's silly to say that Rebecca is Rebecca's sibling, but according to the Prolog statements that we wrote, Rebecca *is* Rebecca's sibling. We didn't write the definition correctly. The Prolog program that we wrote didn't include all of the important facts, and so the Prolog interpreter inferred a fact that we would say is incorrect. That's one of the tricks of writing Prolog, or indeed, of writing good code in *any* programming language. We need to be specific and explicit about what we define and make sure that we don't leave gaps in our code or rely on common sense or intuition. In our code to define sibling, we need to add a clause saying that no one is their own sibling.

logic/family.pl

```
sibling(X, Y) :-
  X \= Y,
  parent(P, X),
  parent(P, Y).

cousin(X, Y) :-
  parent(P, X),
  parent(Q, Y),
  sibling(P, Q).
```

And finally, we can write a rule about cousins. A is a cousin of B if A has a parent that we'll call P, and B has a parent that we'll call Q, and P and Q are siblings:

logic/family.pl

```
cousin(X, Y) :-
  parent(P, X),
  parent(Q, Y),
  sibling(P, Q).
```

```
❮ ?-
⇒  parent(X, rebecca).
❮ X = mark ;
  X = jennifer.
  ?-
⇒ grandparent(X, rebecca).
❮ X = irving ;
  X = paul ;
  X = gail ;
  X = yin.
⇒ cousin(X, rebecca).
❮ X = jocelyn;
  X = jocelyn.
```

How did it figure out that Rebecca and Jocelyn are cousins? It used what we told it: it took the general rule about parents and the specific facts (axioms) that we gave it, and it combined them. Given the fact father(mark, rebecca) and the general statement parent(X, Y) :- father(X, Y), it was able to combine the two and infer the fact that parent(mark, rebecca). Similarly, it used the general rule about what a grandparent is and combined it with both the rules for parents and the specific given facts about who is whose father and mother to produce the new facts about who is whose grandparent. It's also interesting to point out why it sometimes produces the same result more than once. That's because it produced a result each time it inferred that result. So for mark and deb, it produced the result that they're siblings twice—once for their common mother and once for their common father. Then when it was figuring out rebecca's cousins, it used both of those sibling relationships between mark and deb, so it said twice that rebecca and jocelyn are cousins.

What Prolog is doing, internally, is producing proofs. When we give it a predicate with a variable, it takes that and looks for values that it can fill in place of that variable for which it can prove that the predicate is true. Any result produced by a Prolog interpreter is actually a fact that Prolog was able to generate a proof for.

A very important property of Prolog programs to notice is that the proof is entirely symbolic! A Prolog interpreter doesn't know what the atoms represent or what the predicates mean. They're just symbols in a computer's memory. That's true about all logic: inference is a mechanical process. Given a set of atoms and facts and a set of inference rules, a logic can derive proofs without any knowledge of what it all means. The inference process in logic is purely symbolic and can be done with absolutely no clue at all about what the statements that you're proving mean. It's all a mechanical process of working from the premises using the inference rules. Given the right set of premises, you can prove almost any statement; given a choice of both logics and premises, you can prove absolutely any statement.

Computation with Logic

We can use Prolog for much more than just a silly example like family-relationship inferences. In Prolog, using nothing but logical inference, we can implement programs to perform any computation that we could implement in a more traditional programming language. To see how we can do that, we'll look at two examples of how you write more realistic computations in Prolog. First, we'll pull out Peano arithmetic, which we talked about back in 1, *Natural Numbers*, on page 3, and implement it in Prolog. Then we'll take the most widely used sorting algorithm in modern software and look at how it appears in Prolog.

Peano Arithmetic in Prolog

As we saw back toward the beginning of this book, Peano arithmetic is a formal, axiomatic way of defining the natural numbers. It starts by defining zero, and then it defines all of the other natural numbers in terms of a successor operation. Arithmetic is defined structurally by talking about successor relations between numbers.

We can use Prolog to implement natural numbers and their arithmetic operations using the axioms of Peano arithmetic. In fact, Prolog makes it downright easy! Prolog has a very simple mechanism for building data structures that we'll use. In Prolog, any lowercase identifier followed by parens is a *data constructor*. We'll use `z` for zero, and we'll create a data constructor, `s`, to represent the successor of `n`. The data constructor isn't a function. It doesn't *do* anything to its parameters except wrap them. If you're familiar with programming in a language like C++, you can think of a data constructor as being roughly equivalent to `struct S: public NaturalNumber { NaturalNumber* n; };`, except that you don't need to declare the struct type in advance. So when you see `s(n)`, you can think of `s` as being the data type and `s(n)` as meaning the same as `new S(n)`. If `z` is zero, then `s(z)` is one, `s(s(z))` is two, and so on.

logic/number.pl

```
nat(z).
nat(X) :-
  successor(Y, X), nat(Y).
successor(s(A), A).
```

Now we can define arithmetic on our Peano naturals. We can start with addition:

logic/number.pl

```
natadd(A, z, A).
natadd(A, s(B), s(C)) :-
  natadd(A, B, C).
```

The first part, natadd(A, z, A) says that adding zero to any number A gives you A. The second part says that s(C) (the successor to C) is the sum of A and S(B) if C is the sum of A and B.

Let's play with that just a bit.

```
⇒ natadd(s(s(s(z))), s(s(z)), S).
❮ S = s(s(s(s(s(z)))))
```

That's right: 3 + 2 = 5. With that out of the way, we can move on to other ways of using the same predicate.

Unlike most programming languages, in Prolog you don't distinguish between parameters and return values: Prolog will take whichever parameters to a predicate are unbound variables and try to assign values to those variables to make the predicate true. You can invoke a predicate like natadd with all different combinations of bound and unbound parameters:

```
⇒  natadd(s(s(z)), P, S).
❮ P = z,
  S = s(s(z)) ;
  P = s(z),
  S = s(s(s(z))) ;
  P = s(s(z)),
  S = s(s(s(s(z)))) ;
  P = s(s(s(z))),
  S = s(s(s(s(s(z))))) ;
  P = s(s(s(s(z)))),
  S = s(s(s(s(s(s(z)))))) ;
  P = s(s(s(s(s(z))))),
  S = s(s(s(s(s(s(s(z))))))) ;
  P = s(s(s(s(s(s(z)))))),
  S = s(s(s(s(s(s(s(s(z))))))))
```

In essence, I asked the Prolog interpreter, "What are the values of S and P where P + 2 = S? " And Prolog gave me a list of possible answers until I interrupted it. I could also ask it what numbers sum up to a given sum:

```
⇒ natadd(A, B, s(s(s(s(z))))).
❮ A = s(s(s(s(z)))),
  B = z ;
  A = s(s(s(z))),
  B = s(z) ;
  A = B, B = s(s(z)) ;
  A = s(z),
  B = s(s(s(z))) ;
```

```
A = z,
B = s(s(s(s(z)))) ;
false.
```

We can implement multiplication using the same basic pattern as addition.

logic/number.pl

```
product(z, B, z).
product(s(z), B, B).
product(s(A), B, Product) :-
   natadd(B, SubProduct, Product),
   product(A, B, SubProduct).
```

Multiplication is just repeated addition. It's built the same way as addition, but instead of repeatedly invoking successor, we invoke addition. Because of the way that parameters work, our implementation of multiplication is *also* an implementation of division! If we invoke it with the third parameter unbound, then it's multiplication: in product(s(s(z)), s(s(s(z))), P), then P will be bound to 2 times 3. If we invoke it with the first or second parameter unbound, then it's division: in product(s(s(z)), D, s(s(s(s(s(s(z))))))), then D will be bound to 6 divided by 2.

Obviously we wouldn't write real arithmetic this way. This is an extremely inefficient way of doing arithmetic. But it's not an empty example: this general way of decomposing computation into predicates and then expressing computations using recursive predicate definitions is exactly the way that you would implement real programs.

A Quick Prolog Quicksort

Peano numbers are fine, but they're not realistic. No one is going to actually write a real program using Peano numbers. It's great for understanding how the definitions work and it's a fun program to write, but it's not actually useful.

So now we're going to take a look at a real computation. One of the most common, fundamental algorithms that gets used everywhere all the time is quicksort. It's hard to imagine writing a real program in a language that couldn't do a quicksort! At the same time, if you're not familiar with logic programming, it's hard to see how an algorithm like quicksort can be implemented using logical inference!

We're going to solve that little dilemma by looking at a real implementation of quicksort in Prolog and seeing how the algorithm can be described using logic and how inference can implement it.

A Quick Quicksort Refresher

First a refresher, in case you don't know what quicksort is. Quicksort is a *divide and conquer* algorithm for sorting a list of values. The idea of it is pretty simple. You've got a list of values; for simplicity, let's just say that they're numbers. They're all out of order, and we want to get them in order from smallest to largest. How can we do that quickly?

Quicksort picks out one number from the list, called a *pivot*. Then it goes through the list and collects the values that are smaller than the pivot and puts them into one bucket (called Smaller), and it collects the values that are larger than the pivot and puts them into a second bucket (called Larger). Then it sorts the two buckets and concatenates the sorted Smaller, the pivot, and the sorted Larger, and it's done.

For example, let's look at a small list: [4, 2, 7, 8, 3, 1, 5]. We'll take the first element of the list as a pivot, so Pivot=4, Smaller=[2, 3, 1], and Larger=[7, 8, 5]. Then we sort the two buckets, giving us SortedSmaller=[1, 2, 3], and SortedLarger=[5, 7, 8], and our result is [1, 2, 3] + 4 + [5, 7, 8].

Appending Lists: Recursion Over Lists

Let's see how we'd do a quicksort in Prolog. We'll start with the last step: appending the lists together. Actually, that's already implemented in Prolog's standard library, but we'll implement it ourselves because it's a nice introduction to working with lists in Prolog.

logic/sort.pl

```
/* Append(A, B, C) is true if C = A + B */
append([], Result, Result).
append([Head | Tail], Other, [Head | Subappend]) :-
    append(Tail, Other, Subappend).
```

append describes what it means to combine lists. If I say append(A, B, C), I'm asserting that the list C contains the elements of the list A followed by the elements of the list B.

The way that we'd say that in Prolog is a classic recursive definition. There are two cases: a base case and a recursive case. In the base case, we say that the concatenation of the empty list with any other list is that other list.

The recursive case is a bit trickier. It's not really that hard, but it takes a bit of practice to learn how to piece apart a Prolog declaration. But once we translate it into English, it's pretty clear. What it says is this:

1. Suppose I have three lists.

2. The first one can be split into its first element and the rest of it (which we'll call the tail.). Likewise, the last one can be split into its first element and the rest of it.

3. The last list is the concatenation of the first two lists if

    a. The first elements of the first and third lists are the same, and

    b. The rest of the third list is the concatenation of the tail of the first list and the second list.

The key feature of this implementation is that we never say "In order to concatenate two lists, we need to do this, then that, then the other thing." We just describe, in logical terms, what it means for a list to be the concatenation of two other lists.

Let's look at that in practice. Suppose we want to append [1, 2, 3] and [4, 5, 6]:

1. The first list's head is 1, and its tail is [2, 3].

2. The third list is the concatenation of the first two if: the first element of the third list is the same as the head of the first. So if it's the concatenation, it will start with 1.

3. The rest of the third list has to be the concatenation of [2, 3] and [4, 5, 6].

4. So if the third list is the concatenation, then the rest of it has to be [2, 3, 4, 5, 6].

5. So to be true, the third list must be [1, 2, 3, 4, 5, 6].

Partitioning Logically

The next step to building the pieces we need for quicksort is partitioning. To sort, we need to be able to describe what a partitioned list is. We're going to use the same basic trick that we used for concatenation. We're not giving a procedure for how to do it; instead, we're giving a definition of what it means in terms of logic and letting the inference process turn that definition into a procedure.

logic/sort.pl

```
/* partition(A, B, C, D) is true if C and D are lists
 * where A = C + [B] + D
 */
partition(Pivot, [], [], []).
partition(Pivot, [Head | Tail], [Head | Smaller], Bigger) :-
   Head @=< Pivot,
   partition(Pivot, Tail, Smaller, Bigger).

partition(Pivot, [Head | Tail], Smaller, [Head | Bigger]) :-
   Head @> Pivot,
    partition(Pivot, Tail, Smaller, Bigger).
```

This is another example of the recursion pattern that we used in the append predicate.

1. We start with the base case. If the list to partition is empty, then its two partitions—the smaller and the bigger partitions—must also be empty.

2. Now we get to the interesting bit. If the Head of the list to be partitioned is smaller than the Pivot, then the smaller partition must contain that Head. That's why the smaller partition in the predicate declaration here starts with the head.

   The rest of the list is handled by recursion. We partition the Tail and say that Larger must be whatever is partitioned from Tail as larger than the Pivot, and Smaller must be whatever is partitioned from Tail as smaller than the Pivot plus the Head.

3. This is basically the same as the last case, except that the Head is larger than the Pivot.

Sorting

Finally, the sort predicate.

logic/sort.pl

```
/* quicksort(A, B) is true if B contains the same elements as A
 * but in sorted order.
 */
quicksort([], []).
quicksort([Head|Tail], Sorted) :-
  partition(Head, Tail, Smaller, Bigger),
  quicksort(Smaller, SmallerSorted),
  quicksort(Bigger, BiggerSorted),
  append(SmallerSorted, [Head | BiggerSorted], Sorted).
```

Since we already dealt with partition and append, this is pretty easy.

A list Sorted is the sorted form of the input ([Head|Tail]) if—when you partition Tail around Head, sort the two sublists, and then concatenate them—the result is equal to Sorted.

As I have hopefully convinced you, logical inference is extremely powerful. The power of logic isn't limited to simple textbook examples like the family relations we played with! In fact, much like we saw with our implementation of quicksort, any computation that we could implement using any programming language on any computer we can implement using pure logical inference.

If you have any interest in learning more about the logical approach to programming, there are a couple of fantastic books that I recommend. You can learn more about Prolog as a language from *Programming in Prolog: Using the ISO Standard [CM03]*, by William Clocksin and Christopher Mellish. If you want to know more about the way that you use Prolog to do programming based on logical inference, then you can't go wrong with *The Craft of Prolog [O'K09]*, by Richard O'Keefe. I highly recommend picking up one of these two books and spending some time playing with Prolog. It will be time well spent!

15

Temporal Reasoning

The logic that we've looked at so far, first-order predicate logic, is really powerful. You can do a ton of stuff in it. In fact, as we saw in Prolog, if you can do it with a computer, you can do it with first-order predicate logic (FOPL).

But there are some kinds of reasoning that standard predicate logic is really bad at, such as reasoning about time. In predicate logic, if something is true, then it's always true. There's no notion of time, no way for things to happen in sequence. There's no good way to say in predicate logic that I'm not hungry now but I will be later.

For example, in 2010 I worked for Google, and today I work for Foursquare. If I want to be able to capture that, I can't just use a predicate *WorksFor(Mark, Google)*, because that's not true now. Nor can I say *WorksFor(Mark, Foursquare)*, because that wasn't true two years ago. A predicate in FOPL is always true—not just now but in the past as well as in the future.

Of course, if you're clever, you can always find a way to work around limitations. You can work around the problem of change over time using standard predicate logic. One way is to add a time parameter to every predicate. Then instead of saying *WorksFor(Mark, Foursquare)*, I could say *WorksFor(Mark, Foursquare, 2012)*. But then for all of the typical, non-temporal statements in predicate logic, I would need to add universal statements: *∀t: Person(Mark, t)*. It gets very cumbersome very quickly; and worse, it makes using the logic for reasoning painfully awkward.

There's also another problem with predicate logic: there are lots of temporal statements I'd like to make that have a specific temporal structure that I can't express in first-order logic. I'd like to be able to say things like "Eventually I'll be hungry" or "I'll be tired until I get some sleep." Those are two typical statements about temporal properties. They've got common forms that we're familiar with, and it would be really useful to be able to take advantage of those common forms in logical inference. Unfortunately, in first-order predicate logic, even when I've added time terms to every predicate, it's difficult to define a form like *eventually*.

To say something like *eventually* without repeating a complex series of boilerplate statements, we would need to be able to write a predicate that took another predicate as a parameter. And that, by definition, is second-order logic. Switching from first-order to second-order logic creates a huge number of complications. We really don't want to do that.

So if predicate logic is so awkward for reasoning about time, what do we do? We create a new logic. That may sound silly, but it's something we do all the time in math. After all, a logic is a pretty simple formal system, and we can define new ones whenever we want. So we'll just create a new logic, a *temporal* logic, which will make it easy for us to reason about how things change over time.

Statements That Change with Time

Temporal reasoning is really useful. In order to talk about time, logicians have designed many different temporal logics, including CTL, ATL, CTL*, and LTL to name just a few. I'm going to describe the one I'm most familiar with, which is called *computation tree logic*, or CTL. CTL is designed for reasoning about very low-level computations in computer hardware where operations can modify persistent state, like hardware flags. CTL is a very simple logic, which really can't say very much. In fact, CTL may seem unreasonably simple when you see it. But it's really not; CTL is widely used for real-world practical applications.

CTL may be simple, but it's a typical example of the way that you can look at time in a logic. The semantics or meaning of the logic is based on a general idea called Kripke semantics, which is used in many different kinds of logic that need to be able to describe the passage of time. I'll describe the general idea behind the model for temporal logic in this section, but if you want to know more about the idea of Kripke semantics, check out my blog; I wrote a series of posts about intuitionistic logic.[1]

The starting point for CTL is an extremely simple logic called propositional logic. Propositional logic is, basically, FOPL (first-order predicate logic) where the predicates can't take parameters. In a propositional logic, you could make statements like *MarkHasABigNose* and *JimmyDuranteHasABigNose*, but they'd be totally different, unrelated statements. In a propositional logic, you have a finite set of specific statements, and that's it. There are no variables, no quantifiers, no parameters. (There are predicate extensions to CTL, but they make it vastly more complicated, so we'll stick to the simple, basic propositional version.) We can combine the propositions using the standard propositional logical operators: and, or, implication, and negation.

Where it gets interesting is that we also have a set of *temporal quantifiers* that are used to specify the temporal properties of propositional statements. Every statement in CTL has at least two temporal quantifiers. But before we get into them in detail, we need to talk about the basic model of time in CTL.

The idea of the model CTL, as I said earlier, is based on Kripke semantics. Kripke semantics defines a changing system by using a collection of what are called *worlds*. Statements in the logic have a truth binding in specific worlds. Time is a sequence of changes from the world at one moment in time to the world at another moment in time. In CTL's Kripke semantics, we can't say that *P* is true; we can only say *P* is true *in a specific world.*

Each world defines an assignment of truth values to each of the basic propositions. From each world, there's a set of possible *successor worlds*. As time passes, you follow a path through the worlds. In CTL, a world represents a moment

1. http://scientopia.org/blogs/goodmath/2007/03/kripke-semantics-and-models-for-intuitionistic-logic

in time where the truth assignments define what is true at that moment, and the successors to the world represent the possible moments of time that immediately follow it.

The Kripke semantics of CTL effectively give us a *nondeterministic* model of time. From a given moment, there can be more than one possible future, and we have no way of determining which possible future will come true until we reach it. Time becomes a tree of possibilities: from each moment, you could go to any of its successors: each moment spawns a branch for each of its successors, and each path through the tree represents a timeline for a possible future.

CTL gives us two different ways of talking about time in that tree of possible futures; to make a meaningful temporal statement, we need to combine them.

First, if you look at time from any particular moment, there's a collection of possible paths into the future, so you can talk about things in terms of the space of possible futures. You can make statements that begin with things like "In all possible futures…" or "In some possible futures…."

Second, you can talk about the steps along a particular path into the future, about a sequence of worlds that define one specific future. You can make statements about paths like "…will eventually be true." By putting them together, you can produce meaningful temporal statements: "In all possible futures, *X* will always be true"; or "In at least one possible future, *X* will eventually become true."

Every CTL statement uses a pair of *temporal quantifiers* to specify the time in which the statement is true: one *universe quantifier* and one *path quantifier*.

Universe quantifiers are used to make statements that range over all paths forward from a particular moment in time. Path quantifiers are used to make statements that range over all moments of time on a particular timeline-path. As I said, in CTL statements the quantifiers always appear in pairs: a universe quantifier that specifies what set of potential futures you're talking about and a path quantifier that describes the properties of paths within the set quantified by the universe.

There are two universe quantifiers, which correspond to the universal and existential quantifiers from predicate logic.

A *A* is the *all-universe quantifier*. It's used to say that some statement is true *in all possible futures*. No matter what paths you look at, if you follow enough universe transitions, then the statement following the *A* quantifier will become true.

E *E* is the *existential universe quantifier*. It's used to say that there is at least one possible future reachable from the present moment at which the statement will be true.

Next there are path quantifiers. Path quantifiers are similar to universe quantifiers, except that instead of ranging over a set of possible timelines they range over the time-worlds on a specific timeline-path. There are five path quantifiers, which can be divided into three groups:

X (next) The simplest path quantifier is the immediate quantifier, *X* (next). *X* is used to make a statement about the *very next* time-world on this path.

G (global) *G* is the *universal path quantifier*, also called the *global quantifier*. *G* is used to state a fact about every world-moment on the path. Something quantified by *G* is true at the current moment and will *remain* true for all moments on the path.

F (finally) *F* is the *finally path quantifier*, which is used to state a fact about at least one world-moment along a particular timeline-path. If the path described by the *F* quantifier is followed, then the statement will become true in some world on *F*.

Finally, there are temporal relationship quantifiers. These aren't quite quantifiers in the most traditional sense. Most of the time, quantifiers precede statements and either introduce variables or modify the meaning of the statements that follow them. Temporal relationship quantifiers actually *join together* statements in a way that defines a temporal relationship. There are two relationship quantifiers: *strong until* and *weak until*.

U (strong until) *U* is the *strong until* quantifier. When you have a statement *aUb*, it says that *a* is currently true and that eventually it *will* become false, and when it is no longer true, *b* will be true.

W (weak until) *W*, also known as the *weak until quantifier*, is almost the same as *U*. *aWb* also says that *a* is true, and when it becomes false, *b* must be true. The difference is that in *aUb*, eventually *b* must become true, but in *aWb*, *a* might stay true forever; and if it does, *b* might never become true.

It's hard to see quite how all of these quantifiers will work together when you see them all listed out. But once you see them used, they make sense. So we're going to look at a few examples of statements in CTL, and I'll explain what they mean.

Examples: CTL Statements

- *AG.(Mark has a big nose)*: No matter what happens, at every point in time, Mark will always have a big nose. As my kids love to point out, this is an inescapable, unvarying fact.
- *EF.(Joe lost his job)*: It's possible that in some future, Joe will be fired. (To put it formally, there exists a possible timeline where the statement "Joe lost his job" will eventually become true.)
- *A.(Jane does her job well)W(Jane deserves to get fired)*: For all possible futures, Jane does her job well until a time comes when she no longer does her job well, and if/when that happens she'll then deserve to be fired. But this is using *weak eventually*, and so it explicitly includes the possibility that Jane will continue to do her job well forever, and thus she will never deserve to get fired.
- *A.(Mitch is alive)U(Mitch is dead)*: No matter what happens, Mitch will be alive until he dies, and his eventual death is absolutely inevitable.
- *AG.(EF.(I am sick))*: It's always possible that I'll eventually get sick.
- *AG.(EF.(The house is painted blue) ∨ AG.(The house is painted brown))*: In all possible futures, either the house will eventually be painted blue or it will stay brown. It will never be any color other than blue or brown.

What's CTL Good For?

I said that CTL, despite its simplicity, is actually very useful. What's it really good for?

One of the main uses of CTL is something called *model checking*. (I'm not going to go into detail, because the details have easily filled whole books themselves. If you want to know more about model checking with CTL, I recommend *Clarke's textbook [CGP99].*) Model checking is a technique used by both hardware and software engineers to check the correctness of certain temporal aspects of their system. They write a specification of their system in terms of CTL, and then they use an automated tool that compares an actual implementation of a piece of hardware or software to the specification. The system can then verify whether or not the system does what it's supposed to; and if not, it can provide a specific counterexample demonstrating when it will do the wrong thing.

In hardware model checking, you've got a simple piece of hardware, like a single functional unit from a microprocessor. That hardware is basically a complex finite state machine. The way that you can think of it is that the hardware has some set of points where it can have a zero or a one. Each of those points can be represented by a CTL proposition. Then you can describe operations in terms of how outputs are produced from inputs.

For example, if you were looking at a functional unit that implements division, one of the propositions would be "The divide-by-zero flag is set." Then your specification would include statements like *AG.(DivisorIsZero) ⇒ AF.(DivideByZeroFlag)*. That specification says that if the divisor is zero, then eventually the divide-by-zero flag will be set. It does not specify how long it will take: it could take one clock tick in your hardware, or it could take 100. But since the behavior that we care about with that specification should be true regardless of the details of how the divider is implemented, we don't want to specify how many steps, because there can be many different implementations of the specification, which have different precise timing characteristics (think of the difference between a 1990s-era Intel Pentium and a 2012

Intel Core i7; they implement the same instructions for arithmetic, but the hardware is very, very different). The important behavior is that if you try to divide by zero, the appropriate flag bit *will be* set.

The real hardware specifications are a lot more complex than this example, but that gives you the general sense. This is a real-world, common application of CTL: the processor in the computer that I'm typing this on was modeled in CTL.

It's also used in software. Several years ago I worked at IBM. While I was there, I had a friend who did some really fascinating work on using model checking for software. Lots of people had looked at that, because the idea of being able to automatically verify the correctness of software is very attractive. But, sadly, for most software, model checking didn't turn out to be very useful—writing the specifications is hard, and checking them given the amount of state in a typical program is a nightmare! My friend realized that there is a place in software where model checking could be perfect! Modern computing systems use parallel computation and multithreading all the time, and one of the hardest problems is ensuring that all of the parallel threads synchronize properly. The desired synchronization behavior of parallel computation is generally pretty simple and is almost ideally suited for description in a language like CTL. So he worked out a way to use model checking to verify the correctness of the synchronization behaviors of software systems.

Those are the basics of CTL, an extremely simple but extremely useful logic for describing time-based behaviors.

Part V

Sets

Set theory, along with its cousin first-order predicate logic (FOPL), is pretty much the foundation of all modern math. You don't absolutely need set theory, because you can construct math from a lot of different foundations. But the combination of FOPL and axiomatic set theory is currently the dominant approach. Set theory gives us the objects to reason about, and FOPL gives us the ability to do reasoning. The combination of the two gives us math.

There are many kinds of objects that we could use to build math. We could start with numbers or with functions or with a plane full of points. But in modern math we always start with set theory, and not with any of the alternatives! Set theory starts with some of the simplest ideas and extends them in a reasonably straightforward way to encompass the most astonishingly complicated ones. It's truly remarkable that none of set theory's competitors can approach its intuitive simplicity. In this section of the book, we're going to look at what sets are, where they came from, how set theory is defined, and how we can use it to build up other kinds of math.

We'll start by looking at the origins of set theory in the work of Georg Cantor. Cantor's work provides one of the most amazing and counterintuitive results in math. It's a beautiful example of what makes set theory so great. You start with something that seems too simple to be useful, and out of it comes something amazingly profound.

16

Cantor's Diagonalization: Infinity Isn't Just Infinity

Set theory is unavoidable in the world of modern mathematics. Math is taught using sets as the most primitive building block. Starting in kindergarten, children are introduced to mathematical ideas using sets! Since we've always seen it presented that way, it's natural that we think about set theory in terms of foundations. But in fact, when set theory was created, that wasn't its purpose at all. Set theory was created as a tool for exploring the concept of infinity.

Set theory was invented in the nineteenth century by a brilliant German mathematician named Georg Cantor (1845–1918). Cantor was interested in exploring the concept of infinity and, in particular, trying to understand how infinitely large things could be compared. Could there possibly be *multiple* infinities? If there were, how could it make sense for them to have different sizes? The original purpose of set theory was as a tool for answering these questions.

The answers come from Cantor's most well-known result, known as *Cantor's diagonalization*, which showed that there were at least two different sizes of infinity: the size of the set of natural numbers and the size of the set of real numbers. In this chapter, we're going to look at how Cantor defined set theory and used it to produce the proof. But before we can do that, we need to get an idea of what set theory is.

Sets, Naively

What Cantor originally invented is now known as *naive set theory*. In this chapter, we'll start by looking at the basics of set theory using naive set theory roughly the way that Cantor defined it. Naive set theory is easy to understand, but as we'll see in Section 16.3, *Don't Keep It Simple, Stupid,* on page 135, it's got some problems. We'll see how to solve those problems in the next chapter; but for now, we'll stick with the simple stuff.

A *set* is a collection of things. It's a very limited sort of collection where you can only do one thing: ask if an object is in it. You can't talk about which object comes first. You can't even necessarily list all of the objects in the set. The only thing you're guaranteed to really be able to do is ask if specific objects are in it.

The formal meaning of sets is simple and elegant: if an object is a member of a set S, then there's a predicate P_S, where an object o is a member of S (written $o \in S$) if and only if $P_S(o)$ is true. Another way of saying that is that a set S is a collection of things that all share some property, which is the defining property of the set. When you work through the formality of what a property means, that's just another way of saying that there's a predicate. For example, we can talk about the set of natural numbers: the predicate *IsNaturalNumber(n)* defines the set.

Set theory, as we can see even from the first definition, is closely intertwined with first-order predicate logic. In general, the two can form a nicely closed formal system: sets provide objects for the logic to talk about, and logic provides tools for talking about the sets and their objects. That's a big part of why set theory makes such a good basis for mathematics—it's one of the simplest things that we can use to create a semantically meaningful complete logic.

I'm going to run through a quick reminder of the basic notations and concepts of FOPL; for more details, look back at Part IV, *Logic,* on page 77.

In first-order predicate logic, we talk about two kinds of things: *predicates* and *objects*. Objects are the things that we can reason about using the logic; predicates are the things that we use to reason about objects.

A *predicate* is a statement that says something about some object or objects. We'll write predicates as either uppercase letters or as words starting with an uppercase letter (*A*, *B*, *Married*), and we'll write objects in quotes. Every predicate is followed by a list of comma-separated objects (or variables representing objects).

One very important restriction is that *predicates are not objects*. That's why this is called *first-order* predicate logic: you can't use a predicate to make a statement about another predicate. So you can't say something like *Transitive(GreaterThan)*: that's a second-order statement, which isn't expressible in first-order logic.

We can combine logical statements using *and* (written $\wedge$) and *or* ($\vee$). We can negate a statement by prefixing it with *not* (written $\neg$). And we can introduce a variable to a statement using two logical quantifiers: for all possible values ($\forall$), and for at least one value ($\exists$).

When you learned about sets in elementary school, you were probably taught about another group of operations that seemed like primitives. In fact, they aren't really primitive: The only things that we need to define naive set theory is the one definition we gave! All of the other operations can be defined using FOPL and membership. We'll walk through the basic set operations and how to define them.

The basics of set theory give us a small number of simple things that we can say about sets and their members. These also provide a basic set of primitive statements for our FOPL:

Subset

$$S \subseteq T$$

> *S* is a subset of *T*, meaning that all members of *S* are also members of *T*. Subset is really just the set theory version of implication: if *S* is a subset of *T*, then in logic, $S \Rightarrow T$.
>
> For example, let's look at the set *N* of natural numbers and the set N_2 of even natural numbers. Those two sets are defined by the predicates *IsNatural(n)* and *IsEvenNatural(n)*.

When we say that N_2 is a subset of N, what that means is $\forall x$: *IsEvenNatural(x)* $\Rightarrow$ *IsNatural(x)*.

Set Union

$$A \cup B$$

Union combines two sets: the members of the union are all of the objects that are members of either set. Here it is in formal notation:

$$x \in (A \cup B) \lneqq x \in A \vee x \in B$$

The formal definition also tells you what union means in terms of logic: union is the logical *or* of two predicates.

For example, if we have the set of even naturals and the set of odd naturals, their union is the set of objects that are either even naturals or odd naturals: an object x is in the union *(EvenNatural* $\cup$ *OddNatural)* if either *IsEvenNatural(x)* or *IsOddNatural(x)*.

Set Intersection

$$A \cap B$$

The intersection of two sets is the set of objects that are members of both sets. Here it is presented formally:

$$x \in A \cap B \lneqq x \in A \wedge x \in B$$

As you can see from the definition, intersection is the set equivalent of logical *and*.

For example, *EvenNatural* $\cap$ *OddNatural* is the set of numbers x where *EvenNatural(x)* $\wedge$ *OddNatural(x)*. Since there are no numbers that are both even and odd, that means that the intersection is empty.

Cartesian Product

$$A \times B$$
$$(x, y) \in A \times B \lneqq x \in A \wedge y \in B$$

Finally, within the most basic set operations, there's one called the *Cartesian product*. This one seems a bit weird, but it's really pretty fundamental. It's got two purposes: first, in practical terms, it's the operation that lets us create ordered pairs, which are the basis of how we can create virtually everything that we want using sets. In purely theoretical terms, it's the way that set theory expresses the concept of a predicate that takes more than one parameter. The Cartesian product of two sets S and T consists of a set of *pairs,* where each pair consists of one element from each of the two sets.

For example, in 12, *Mr. Spock Is Not Logical,* on page 79, we defined a predicate *Parent(x, y),* which meant that x is a parent of y. In set theory terms, *Parent* is a set of *pairs* of people. So *Parent* is a subset of the values from the Cartesian product of the set of people with itself. *(Mark, Rebecca)* $\in$ *Parent,* and *Parent* is a predicate on the set *Parent* × *Parent.*

That's really the heart of set theory: set membership and the linkage with predicate logic. It's almost unbelievably simple, which is why it's considered so darned attractive by mathematicians. It's hard to imagine how you could start with something simpler.

Now that you understand how simple the basic concept of a set is, we'll move on and see just how deep and profound that simple concept can be by taking a look at Cantor's diagonalization.

Cantor's Diagonalization

The original motivation behind the ideas that ended up growing into set theory was Cantor's recognition of the fact that there's a difference between the size of the set of natural numbers and the size of the set of real numbers. They're both infinite, but they're not the same!

Cantor's original idea was to abstract away the details of numbers. Normally when we think of numbers, we think of them as being things that we can do arithmetic with, things that can be compared and manipulated in all sorts of ways. Cantor said that for understanding how many numbers there were, none of those properties or arithmetic operations were needed. The only thing that mattered was that a kind of number like the natural numbers was a collection of objects. What mattered is which objects were parts of which collection. He called this kind of collection a *set.*

Using sets allowed him to invent a new way of defining a way of measuring size that didn't involve counting. He said that if you can take two sets and show how to create a mapping from every element of one set to exactly one element of the other set, and if this mapping didn't miss any elements of either set (a *one-to-one mapping* between the two sets), then those two sets are the same size. If there is no way to make a one-to-one mapping without leaving out elements of one set, then the set with extra elements is the *larger* of the two sets.

For example, if you take the set *{1, 2, 3}*, and the set *{4, 5, 6}*, you can create several different one-to-one mappings between the two sets: for example, $\{1 \Rightarrow 4, 2 \Rightarrow 5, 3 \Rightarrow 6\}$, or $\{1 \Rightarrow 5, 2 \Rightarrow 6, 3 \Rightarrow 4\}$. The two sets are the same size, because there is a one-to-one mapping between them.

In contrast, if you look at the sets *{1, 2, 3, 4}* and *{a, b, c}*, there's no way that you can do a one-to-one mapping without leaving out one element of the first set; therefore, the first set is larger than the second.

This is cute for small, finite sets like these, but it's not exactly profound. Creating one-to-one mappings between finite sets is laborious, and it always produces the same results as just counting the number of elements in each set and comparing the counts. What's interesting about Cantor's method of using mappings to compare the sizes of sets is that mappings can allow you to compare the sizes of infinitely large sets, which you *can't* count!

For example, let's look at the set of natural numbers (N) and the set of even natural numbers (N_2). They're both infinite sets. Are they the same size? Intuitively, people come up with two different answers for whether one is larger than the other.

1. Some people say they're both infinite, and therefore they must be the same size.

2. Other people say that the even naturals must be half the size of the naturals, because it skips every other element of the naturals. Since it's skipping, it's leaving out elements of the naturals, so it must be smaller.

Which is right? According to Cantor, both are wrong. Or rather, the second one is completely wrong, and the first is right for the wrong reason.

Cantor says that you can create a one-to-one mapping between the two:

$$\{(x \gg y) : x, y \in N, y = 2 \times x\}$$

Since there's a one-to-one mapping, that means that they're the same size—they're not the same size because they're both infinite, but rather because there is a one-to-one mapping between the elements of the set of natural numbers and the elements of the set of even natural numbers This shows us that some infinitely large sets are the same size as some other infinitely large sets. But are there infinitely large sets whose sizes are *different*? That's Cantor's famous result, which we're going to look at.

Cantor showed that the set of real numbers is larger than the set of natural numbers. This is a very surprising result. It's one that people struggle with because it *seems* wrong. If something is infinitely large, how can it be smaller than something else? Even today, almost 150 years after Cantor first published it, this result is *still* the source of much controversy (see, for example, *this famous summary [Hod98].*) Cantor's proof shows that no matter what you do, you can't create a one-to-one mapping between the naturals and the reals without missing some of the reals; and therefore, the set of real numbers is larger than the set of naturals.

Cantor showed that every mapping from the naturals to the reals *must* miss at least one real number. The way he did that is by using something called a *constructive proof*. This proof contains a procedure, called a *diagonalization*, that takes a purported one-to-one mapping from the naturals to the reals and generates a real number that is missed by the mapping. It doesn't matter what mapping you use: given *any* one-to-one mapping, it will produce a real number that isn't in the mapping.

We're going to go through that procedure. In fact, we're going to show something even stronger than what Cantor originally did. We're going to show that there are more real numbers between zero and one than there are natural numbers!

Cantor's proof is written as a basic proof by contradiction. It starts by saying "Suppose that there is a one-to-one mapping from the natural numbers to the real numbers between zero and one." Then it shows how to take that supposed mapping and use it to construct a real number that is missed by the mapping.

Example: Prove that there are more real numbers between 0 and 1 than there are natural numbers.

1. Suppose that we can create a one-to-one correspondence between the natural numbers and the reals between 0 and 1. What that would mean is that there would be a total one-to-one function *R* from the natural numbers to the reals. Then we could create a complete list of all of the real numbers: *R(0), R(1), R(2), ...*

2. If we could do that, then we could also create another function, *D* (for digit), where *D(x,y)* returns the *y*th digit of the decimal expansion of *R(x)*. The *D* that we just created is effectively a table where every row is a real number and every column is a digit position in the decimal expansion of a real number. *D(x,3)* is the third digit of the binary expansion of *x*.

   For example, if *x* = *3/8*, then the decimal expansion of *x* is 0.125. Then *D(3/8,1)* = *1*, *D(3/8,2)* = *2*, *D(3/8,3)* = *5*, *D(3/8,4)* = *0, ...*

3. Now here comes the nifty part. Take the table for *D* and start walking down the diagonal. We're going to go down the table looking at *D(1,1), D(2,2), D(3,3),* and so on. And as we walk down that diagonal, we're going to write down digits. If the *D(i, i)* is 1, we'll write a 6. If it's 2, we'll put 7; 3, we'll put 8; 4 ⇒ 9; 5 ⇒ 0; 6 ⇒ 1; 7 ⇒ 2; 8 ⇒ 3; 9 ⇒ 4; and 0 ⇒ 5.

4. The result that we get is a series of digits; that is, a decimal expansion of some number. Let's call that number *T*. *T* is *different* from every row in D in at least one digit—for the *i*th row, T is different at digit *i*. There's no *x* where *R(x)* = *T*.

But T is clearly a real number between 0 and 1: the mapping can't possibly work. And since we didn't specify the structure of the mapping, but just assumed that there was one, that means that there's no possible mapping that will work. This construction will always create a counterexample showing that the mapping is incomplete.

5. Therefore, the set of all real numbers between 0 and 1 is *strictly larger* than the set of all natural numbers.

That's Cantor's diagonalization, the argument that put set theory on the map.

Don't Keep It Simple, Stupid

There's an old mantra among engineers called the KISS principle. KISS stands for "Keep it simple, stupid!" The idea is that when you're building something useful, you should make it as simple as possible. The more moving parts something has, the more complicated corners it has, the more likely it is that an error will slip by.

Looked at from that perspective, naive set theory looks great. It's so beautifully simple. What I wrote in the last section was the entire basis of naive set theory. It looks like you don't need any more than that!

Unfortunately, set theory in practice needs to be a lot more complicated. In the next section, we'll look at an axiomatization of set theory, and yeah, it's going be a whole lot more complicated than what we did here! Why can't we stick with the KISS principle, use naive set theory, and skip that hairy stuff?

The sad answer is, naive set theory doesn't work.

In naive set theory, *any* predicate defines a set. There's a collection of mathematical objects that we're reasoning about, and from those, we can form sets. The sets themselves are also objects that we can reason about. We did that a bit already by defining things like subsets, because a subset is a relation between sets.

By reasoning about properties of sets and relations between sets, we can define sets of sets. That's important, because sets of sets are at the heart of a lot of the things that we do with set theory. As we'll see later, Cantor came up with a way of modeling numbers using sets where each number is a particular kind of structured set.

If we can define sets of sets, then using the same mechanism, we can create infinitely large sets of sets, like "the set of sets with infinite cardinality," also known as the set of infinite sets. How many sets are in there? It's clearly infinite. Why? Here's a sketch: if I take the set of natural numbers, it's infinite. If I remove the number 1 from it, it's still infinite. So now I have two infinite sets: the natural numbers, and the natural numbers omitting 1. I can do the same for every natural number, which results in an infinite number of infinite sets. So the set of sets with infinite cardinalities clearly has infinite cardinality! Therefore, it's a member *of itself*!

If I can define sets that contain themselves, then I can write a predicate about self-inclusion and end up defining things like the set of all sets that include themselves. This is where trouble starts to crop up: if I take that set and examine it, does it include itself? It turns out that there are *two sets* that match that predicate! There's one set of all sets that include themselves that includes itself, and there's another set of all sets that include themselves that does *not* include itself.

A predicate that *appears* to be a proper, formal, unambiguous statement in FOPL turns out to be ambiguous when used to define a set. That's not fatal, but it's a sign that there's something funny happening that we should be concerned about.

But now, we get to the trick. If I can define the set of all sets that contain themselves, I can also define the set of all sets that do *not* contain themselves.

And that's the heart of the problem, called *Russell's paradox*. Take the set of all sets that do *not* include themselves. Does it include itself?

Suppose it does. If it does, then by its definition, it *cannot* be a member of itself.

So suppose it doesn't. Then by its definition, it *must be* a member of itself.

We're trapped. No matter what we do, we've got a contradiction. And in math, that's deadly. A formal system that allows us to derive a contradiction is completely useless. One error like that, allowing us to derive just one contradiction, means that every result we ever discovered or proved in the system is worthless! If there's a single contradiction possible anywhere in the system, then every statement—whether genuinely true or false—is provable in that system!

Unfortunately, this is pretty deeply embedded in the structure of naive set theory. Naive set theory says that *any* predicate defines a set, but we can define predicates for which there is no valid model, for which there is no possible set that consistently matches the predicate. By allowing this kind of inconsistency, naive set theory itself is inconsistent, and so naive set theory needs to be discarded. What we need to do to save set theory at all is build it a better basis. That basis should allow us to do all of the simple stuff that we do in naive set theory, but do it without permitting contradictions. In the next section, we'll look at one version of that, called Zermelo-Frankel set theory, that defines set theory using a set of strong axioms and manages to avoid these problems while preserving what makes set theory valuable and beautiful.

17

Axiomatic Set Theory: Keep the Good, Dump the Bad

In the last chapter, we saw the basics of naive set theory the way that Cantor defined it. Naive set theory seems wonderful at first because of its brilliant combination of simplicity and depth. Unfortunately, that simplicity comes at great cost: it allows you to create logically inconsistent self-referential sets.

Happily, the great mathematicians of the early twentieth century weren't willing to give up on set theory. The devotion of mathematicians to preserving set theory is best summed up by the words of one of the most brilliant mathematicians in the history of math, David Hilbert (1862–1943): "No one shall expel us from the Paradise that Cantor has created." This devotion led to an effort to reformulate the foundations of set theory in a way that preserved as much of the elegance and power of set theory as possible while eliminating the inconsistencies. The result is what's known as *axiomatic set theory*.

Axiomatic set theory builds up set theory from a set of fundamental initial rules, called *axioms*. We're going to take a look at a set of axioms that produce a consistent form of set theory. There are several different ways of formulating set theory axiomatically, which all produce roughly the same result. We're going to look at the most common version, called the *Zermelo-Frankel set theory with choice*, commonly abbreviated as ZFC. The ZFC axiomatization consists of a set of ten axioms, which we'll look at. In particular, the last axiom, called *the axiom of choice*, is still the focus of controversy today, more than 100 years after it was proposed, and we'll take a look at why it still inspires so much emotion among mathematicians!

There are other choices for axiomatic set theory. Most well-known is an extension of ZFC called *NBG set theory*. We'll say a bit about NBG, but our focus will be ZFC theory.

The Axioms of ZFC Set Theory

In this section, we're going to walk through the process of creating a *sound* set theory, a version of set theory that preserves the intuition and simplicity of naive set theory but won't fall into the trap of inconsistency.

Keep in mind as we go that the point of what we're doing is to produce something that is a foundation. As a foundation, it cannot depend on *anything* except for first-order predicate logic and the axioms themselves. Until we can construct them using the axioms, there are no numbers, no points, no functions! We can't assume that anything exists until we show how it can be constructed using these axioms.

What do we need to create a sound version of set theory? In naive set theory, we started by saying that a set is defined by its members. We'll start the same way in axiomatic set theory. The property that a set is defined solely by its members is called *extensionality*.

The Axiom of Extensionality

$$\forall A, B : A = B \lneqq (\forall C : C \in A \Rightarrow C \in B)$$

> This is a formal way of saying that a set is described by its members: two sets are equivalent if and only if they contain the same members. The name *extensionality* is mathematical terminology for saying that things are equal if they behave the same way. Since the only real behavior of sets is testing if an object is a member of a set, two sets are equal if they always have the same answer when you ask if an object is a member.

The axiom of extensionality does two things: it defines what a set *is* by defining its behavior, and it defines how to compare two sets. It doesn't say anything about whether any sets actually exist or how we can define new sets.

The next step then, is to add axioms that let us start building steps. We follow the inductive pattern that we saw back in Section 1.1, *The Naturals, Axiomatically Speaking*, on page 4: there's a base case and an inductive case. The base case says that there is a set containing no elements. The inductive cases tell us how we can build other sets using the empty set.

The Empty-Set Axiom

$$\exists \emptyset : \forall x : x \neg \in \emptyset$$

The empty set axiom gives us the base case of all sets: it says that there exists a set, the empty set, which contains no members. This gives us a starting point: an initial value that we can use to build other sets. It does a bit more than that: by telling us that there is an empty set, it tells us that the set of values that exists given this axiom is *the set containing the empty set.*

The Axiom of Pairing

$$\forall A, B : (\exists C : (\forall D : D \in C \Rightarrow (D = A \vee D = B)))$$

Pairing gives us one way of creating new sets. If we have any two objects, then we can create a new set consisting of exactly those two objects. Before pairing, we could have the empty set ($\emptyset$) and the set containing the empty set ($\{\emptyset\}$), but we had no way of creating a set containing both the empty set *and* the set containing the empty set ($\{\emptyset, \{\emptyset\}\}$). In logical form, pairing just says that we can build a two-item set by enumeration: given any two sets A and B, there's a set C containing *only* A and B as members.

The Axiom of Union

$$\forall A : (\exists B : (\forall C : C \in B \lneq (\exists D : C \in D \wedge D \in A)))$$

The axiom of union is the best friend of the axiom of pairing. With the two of them together, we can create any finite set we want by enumerating its members. What it says formally is that given any two sets, their union is a set. The notation is complicated, because we haven't defined the union operation yet, but all that it really says is that we can define the union operation and create new sets by union. By using the axiom of pairing, we can pick specific elements that we want in a set; and by using the axiom of union, we can chain together multiple sets created by pairing. Thus with the two of them, we can provide a specific list of elements, and there will be a set containing exactly those elements and no others.

The four axioms that we've seen so far give us the ability to create any finite set we want. But finite sets aren't enough: we want things like Cantor's diagonalization to work with our new set theory. We need some way of creating more than just finite sets. This is a crucial point for our new set theory: infinite sets are where Cantor's naive set theory ran into trouble. Any mechanism that lets us create infinite sets must be designed very carefully to ensure that it *cannot* create self-referential sets.

We'll start by creating a single, canonical infinite set. We know that this canonical infinite set is well behaved. Then we'll use that set as a prototype: every infinite set will be derived from the master infinite set and therefore will not contain anything that could allow it to become a problem.

The Axiom of Infinity

$$\exists N: \emptyset \in N \land (\forall x : x \in N \Rightarrow x \cup \{x\} \in N)$$

The axiom of infinity is the most difficult axiom we've seen so far, because it's introducing a genuinely difficult idea: it's saying that we can have sets containing an infinite number of members! Instead of just asserting that we can have infinite sets, it says that there's a *specific way* to create infinite sets, and that all infinite sets must either be created this way or derived from another infinite set created this way.

The prototypical infinite set is defined by the axiom as the set that

1. Contains the empty set as a member, and

2. For each of its members x, also contains the singleton set *{ x }* containing x. That means that if we

followed the formal statement and named the set N, N contains $\varnothing$, $\{\varnothing\}$, $\{\{\varnothing\}\}$, $\{\{\{\varnothing\}\}\}$, and so on.

The axiom really does two things. First, it does what we've already talked about: it gives us the prototypical infinite set. But second, it defines that infinite set in a very specific way. We could have defined a canonical infinite set in many different ways! The reason that we defined it this way is because this specific construction is derived directly from the way that Peano numbers can be constructed. This means basically that the Peano numbers *are* the canonical infinite set.

The Meta-Axiom of Specification

$$\forall A : \exists B : \forall C : C \in B \Rightarrow C \in A \wedge P(C)$$

For finite-size sets, the axiom of the empty set gave us a prototypical finite-size set, and then the axioms of pairing and union let us use that initial finite-size set to create the other sets. For infinite sets, the axiom of infinity gave us a prototypical infinite set, and now the meta-axiom of specification will allow us to create as many other infinite sets as we want using logical predicates.

It's called a meta-axiom because it's got a big problem. What we *want* it to say is this: given any predicate P, you can take any set A and select out the elements of A for which P is true, and the resulting collection of elements is a set. Less formally, we want it to say that you can create a set of objects that share a particular property by starting with a larger set (like our prototypical infinite set) and then select the subset that has the desired property.

That's what we *want* to be able to say with a single axiom. Unfortunately we can't say that. A statement that's true *for any predicate P* is impossible to write in first-order logic. To get around this problem, the people who designed ZFC set theory did the only thing that they could: they *cheated* and said it's not really a second-order axiom but a schema for what's actually an infinite set of axioms. For every predicate P, there's another instantiation of the axiom of specification that says you can use that predicate P to define subsets of any set.

Nowhere in the axiom does it say that you can only use it to select elements from an infinite set. You can use the axiom of specification to find subsets of finite sets with a common property. But for finite sets, you don't *need* to use specification: you can just manually enumerate the elements that you want. If you had the set $A = \{1, 2, 3, 4\}$, you could say, "The elements of A that are even," or you could just say, "The set {2, 4}." With infinite sets, you need to use specification to define a set like the set of even natural numbers.

The Powerset Construction Axiom

$$\forall A : \exists B : \forall C \subseteq A : C \in B$$

This is a nice, easy one, but it's very important. The axiom of infinity gave us a prototypical infinite set. The axiom of specification gave us a way of creating other infinite sets by selecting elements from an infinite set. Using the two together, we can create infinite sets that are *subsets* of the prototype. However, that's just not enough. We know that if our new set theory works, then the set of real numbers will be *larger* than the set of natural numbers, and we know that the prototypical infinite set is exactly the same size as the set of naturals. Without some way of creating a set larger than that, we won't be able to represent the real numbers using sets! The powerset axiom gives us a way around that: it says that for any set A, the collection of all of the possible subsets of A (called the *powerset* of A) is also a set. With this, the axiom of infinity, and the axiom of specification, we can create a whole universe of infinite sets!

The axiom of the powerset is dangerous. As soon as we can create a set *larger* than the set of natural numbers, we're treading on thin ice: that's where the self-referential sets that caused us so much pain live. Explaining exactly why is too complex for this book, but until we had infinite sets larger than the set of natural numbers, we couldn't build a paradoxical self-referential set. With the powersets, though, we can, so to keep our theory sound we need to put up a firewall to prevent that.

The Foundation Axiom

$$\forall A \neq \emptyset : \exists B \in A : A \cap B = \emptyset$$

Every set A contains some member B that is a set completely disjoint from A.

Understanding the point of this one requires a bit of brain-bending. It's making it impossible to build certain kinds of self-referential sets that would lead to inconsistency. You can't create a set that only contains itself, and in fact, every set like the problematic set of all sets that do not contain themselves simply can't be expressed in a paradoxical way without violating this axiom. Therefore they aren't sets, we can't create them, we aren't talking about them when we reason about sets, and thus they aren't a problem. It may feel a bit "handwavy," but what we're really doing is forbidding those sets in the least restrictive way that we can.

We're almost done with our new set theory! We can create finite sets and infinite sets. We can create sets of whatever size we want, from finite to infinite and beyond! We can define sets using predicates. And we can do all of that without causing any inconsistencies. What's left?

Two things remain. The first is easy; the second is one of the most difficult ideas that I've ever encountered. The easy one is just a formal way of saying how we can define functions using sets and predicates. The other one...well, we'll get to that.

The Meta-Axiom of Replacement

$$\forall A : \exists B : \forall y : y \in B \Rightarrow \exists x \in A : y = F(x)$$

Replacement is another meta-axiom, a stand-in for an infinite number of real axioms. It says that you can define functions in terms of predicates: *P(x,y)* is a function as long as the set defined by it has the key property of a function: each value in its domain is mapped to one value in its range. We can't say it that simply, because we haven't defined functions, domains, or ranges.

We need this because when we talk about a typical function, in formal terms, the function is defined logically using a predicate. For example, when I define the function $f(x) = x^2$, what I'm actually saying is that in formal terms there is a set of pairs (a, b) for which a predicate that means "b is the square of a" is true. That runs into exactly the same problem that we had in the meta-axiom of specification: we want to be able to say that for any function-ish predicate, we can define the function in terms of sets. Since we can't say anything about *all* predicates, we have to explicitly say that you can do this, and we have to say it as a meta-axiom that stands in for an infinite number of real axioms, one for each function-ish predicate.

Hang on tight; here's where it gets hairy! The axiom of choice is still a subject of debate among mathematicians. It's a very tricky concept. It's extremely subtle, and it's difficult to explain why it's needed in intuitive terms. It's really a counterpart for the axiom of replacement. Replacement provided a capability that we need to define functions, because we knew that we were going to want to be able to talk about functions, and without defining the capability with an axiom, it wouldn't be possible. Similarly, the axiom of choice provides a capability that we know we're going to need for subjects like topology, so we need to add it as an axiom.

The Axiom of Choice I'm not going to try to do this in pure notation; it's just too hard to understand.

- Let X be a set whose members are all non-empty.
- There exists a function f from X to the union of the members of X, such that for each member $x \in X$, $f(x) \in X$.
- f is called a *choice function*.

What does that mean? It's not an easy thing to explain. Roughly, it means that sets have a certain kind of structure that makes it possible to say that *there exists* a consistent and yet arbitrary mechanism for selecting things from that set, even when the set is infinitely large, even though we *can't* create that mechanism or say how it could possibly work.

That's it: we have our new set theory, which is defined by this collection of axioms. This new version of set theory has everything that we love about set theory. It's all stated formally, which makes it more complicated to read, but the underlying concepts remain nearly the same as the intuitive ideas of naive set theory. The careful axiomatic basis preserves all of that, but it guarantees that we're safe from inconsistencies: the problems that broke naive set theory are prevented by the careful design of the axioms.

But we're not done yet. I've said that the axiom of choice is controversial and difficult, but I haven't really explained why. So now we're going to take a more in-depth look at why the axiom of choice is strange and why we need it even though it causes so much weirdness.

The Insanity of Choice

The only one of the axioms that's really difficult to understand is the axiom of choice. And that itself makes sense, because even among mathematicians, the axiom of choice is downright controversial in some circles. There's a bunch of variations of it, or different ways of ultimately saying the same thing. But they all end up with the same problems. And those problems are whoppers. But dumping choice also has some serious problems.

To understand just what makes choice such a problem, it's easiest to use a variation of it. To be clear about what I mean by a variation: it's not just a restatement of the axiom of choice in different language, but a statement that ends up proving the same things. For example, we're going to look at a variation called the well-ordering theorem. If we keep the axiom of choice, then the well-ordering theorem is provable using the ZFC axioms. If we don't keep the axiom of choice but replace it with the well-ordering theorem, then the axiom of choice is provable using ZFC set theory with well-ordering.

The easiest variation of the axiom of choice is called the well-ordering theorem. It's a great demonstration of the insanity of choice because on the face of it, it sounds pretty easy and obvious. But when you think it through, you realize that it's absolutely crazy. It says that for every set *S*, including infinite sets, it's possible to define a *well-ordering* on *S*. A well-ordering on *S* means that for every subset of *S*, the subset has a unique *smallest element*.

Obvious, right? If we have a less-than operation, then there must be a smallest element, right? It seems to make sense. But it implies something absolutely crazy: the well-ordering theorem says that there's a single, unique value that is the *smallest* positive real number larger! Well-ordering says that for every real number *r*, there is a single, unique real number that comes *next*. After all, I can take the set of real numbers greater than or equal to 0—that's obviously permitted by the axiom of specification. The minimum element of that set is 0. Using the axiom of specification again, I can create a subset of that that doesn't include 0. That new set *still* has a smallest element: the smallest real number greater than 0.

That's ridiculous, isn't it? We know that for every pair of real numbers, there's an infinite number of real numbers that fit between them. It *looks* like it's a major inconsistency, as bad as the inconsistent sets in ZFC. The usual argument is that it's not inconsistent: it's not a problem because we can't ever get enough of a grip on that number to be able to *do* anything with it that exploits its value in a way that creates a definite inconsistency.

Another way of demonstrating the weird implications of choice is one of the weirdest things in math. It's called the *Banach-Tarski paradox*. Banach-Tarski says that you can take a sphere and slice it into six pieces. Then you can take those same six pieces, put them back together—without bending or folding—as *two* identical spheres, each exactly the same size as the original sphere! Or you can take those six pieces and reassemble them into a new sphere that is one thousand times larger than the original sphere!

When you look at it in detail, Banach-Tarski looks like it's downright crazy. It's not as bad as it seems. It *looks* wrong, but it does actually work out in the end. Banach-Tarski isn't inconsistent: it just demonstrates that infinity is a really weird idea. When we talk about things like the volume of a sphere, we're using something called *measure theory*: the volume of a shape is defined by a measure. What's happening in Banach-Tarski is that we're using the axiom of choice to create an *infinitely complex* slice. At infinite complexity, the volume and surface area of a shape *can't* be measured. We've pushed measure theory beyond its limits. When we put the slices back together, the infinitely complex cuts match up, and we end up with a figure that can once again be measured. But because it passed through an indeterminate phase, it doesn't have to be consistent before and after. It works out because ultimately each sphere contains an infinite number of points: if we divide the infinite set in half, each half still has the same number of points, which means that there are obviously enough points to fill the two new spheres. In set theory, that's not inconsistent. In measure theory it is, but measure theory doesn't allow that kind of infinitely complex slice, which is why the volume is unmeasurable.

The axiom of choice seems to cause problems in all of its forms. Why do we insist on keeping it? The easiest way to answer that is to look at yet another way of restating it. The axiom of choice is equivalent to this statement: the Cartesian product of a collection of non-empty sets is not empty. That's such a simple necessary statement: if we couldn't even show that the product of non-empty sets isn't empty, how could we possibly do *anything* with our new set theory?

What the axiom of choice ultimately says is that we can create a function that chooses from a set of indistinguishable things. It doesn't matter what the choice is. It just matters that in some theoretical sense, there is a function that can choose. Bertrand Russell (1872–1970), who was one of the mathematicians who helped establish ZFC set theory as the dominant foundation of mathematics, explained why the axiom of choice was needed by saying, "To choose one sock from each of infinitely many pairs of socks requires the Axiom of Choice, but for shoes the Axiom is not needed." With shoes, the two members of the pair are different, so we can easily design a function that distinguishes between them. The two members of a pair of socks are exactly the same, so we can't say how to distinguish between them. The axiom of choice says that the two can be distinguished, even if we can't imagine how.

Choice is uncomfortable. It seems crazy, and it seems inconsistent. But no one has been able to definitively show that it's *wrong*, just that it's uncomfortable. There are some holdouts that refuse to accept it, but for most of us, the cost of dropping choice is too high. Too much of the math, from basic number theory to topology and calculus, depend on it. It produces results that *feel* wrong, but under the hood, those wrong-seeming results are explainable. At the end of the day, it *works*, so we stick with it.

Why?

There we are. That's it: set theory in a nutshell. You can derive pretty much all of mathematics from those ten axioms plus simple first-order predicate logic. The integers fall out pretty naturally from the axiom of infinity; once you've got the integers, you can use the axiom of pairing to create the rationals; once you've got the rationals, you can use these axioms to derive Dedekind cuts to get the reals; once you've got the reals, you can use the axiom of replacement to get the transfinites. It just all flows out from these ten rules.

The amazing thing is that they're not even particularly hard! The axioms make sense: the reason why each is needed is clear, and what each one means is clear. It doesn't strain our brains to understand them! It took some real genius to derive these rules, figuring out how to draw down the entirety of set theory into ten rules while preventing problems like Russell's paradox is an astonishingly difficult task. But once a couple of geniuses did that for us, the rest of us dummies are in great shape. We don't need to be able to derive them; we just need to understand them.

The point of the entire exercise of defining set theory axiomatically was to avoid the inconsistency of naive set theory. The way that our axioms did that is by hard-wiring constraints into the basic definitions of sets and the fundamental set operations. The axioms don't just say, "A set is a collection of stuff"; they constrain what a set is. They don't just say, "If you can write a predicate that selects the members, that predicate defines a set"; they provide a constrained mechanism by which you can define *valid* sets using predicates. With those constraints in place, we have a consistent, well-defined foundational theory that we can use to build the rest of math!

18

Models: Using Sets as the LEGOs of the Math World

Mathematicians like to say that they can re-create all of mathematics using set theory as a basis. What does that even mean?

Sets are amazingly flexible. With the basic structure given to us by sets in ZFC set theory, we can *build* anything. They're basically a lot like the mathematical version of a kid's LEGO set: they're easy to put together in a lot of different ways, so you can use them to build whatever you want. You can pick pretty much any field of math and build the objects that you need using sets.

Suppose we wanted to build a new mathematical system, like topology. A simple way of describing topology is that it's a way of studying the shape of a surface by looking at which points on that surface are close to which other points. The axioms of topology define things like what a *point* is, what a *shape* is, and what it means for one point to be *close to* another point.

We could build the math of topology starting with absolutely nothing, the way we did with sets. To do that, we'd need to start with a lot of very basic axioms to answer the basic questions: What's the simplest topological space? How can we use that to build more complex spaces? How can we use logic to make statements about spaces? And so on. It would be very difficult, and much of it would retrace the steps that we already followed in defining ZFC. Using set theory makes the process much easier: we can just build a *model* that shows how to build the basic objects of topology (points and surfaces) in terms of sets. Then we can show how the axioms of topology are provable using the model along with the axioms of ZFC.

As with so many ideas in math, it's easiest to understand what I mean by building a model by looking at a specific example. For our example, we'll go back to the original purpose of set theory and use it to build a set-based model of the ordinal and cardinal numbers.

Building Natural Numbers

All the way back in 1, *Natural Numbers,* on page 3, we defined the natural numbers axiomatically. In most of math, we define objects and their behaviors using axiomatic definitions like the rules of Peano arithmetic. One of the more subtle problems in math is that if you want to define a new mathematic construct, it's not enough to just define a collection of axioms. A collection of axioms is a logical definition of a kind of object and the way that that kind of object works, but the axioms don't actually show that an object that fits the definition exists or that it even makes sense to create an object that fits the definition. To make the axiomatic definition work, you need to show that it's possible for the objects described in the definition to exist by showing how you can construct a set of objects that fit the axioms. That set of objects is called *a model* of the axioms. That's where sets come into play: sets provide an ideal framework for building mathematical objects to use in the model for pretty much any axiomatic definition.

We're going to create a model of the natural numbers. How do we do that without assuming that numbers already exist? We look back to the creator of set theory. Back in the nineteenth century, Georg Cantor did something that many mathematicians have tried to do. He tried to find a simple, minimal basis on which he could build a new version of mathematics. What he came up with is what became set theory. Before Cantor's work, the basic ideas of sets had been used in math for thousands of years, but no one had ever put it on a formal basis. Cantor changed the face of math forever by doing exactly that. He didn't get it completely right, but the problems with his ideas about set theory were fixed by ZFC. Without Cantor's pioneering work showing the value of the formal concept of sets, ZFC would never have been developed. Cantor's first demonstration of the power of sets was using set concepts to build a model of numbers, and that model holds up beautifully under ZFC.

Cantor formalized sets as part of his study of the roots of number theory. He wanted to start with the simple set notion and then build a model of numbers. So that's exactly what we're going to do.

First we need to define the objects that we're going to talk about. That's the set of natural numbers. We've already seen the basic construction in the axiom of infinity. The set of natural numbers starts with 0, which we represent as $\emptyset$. For each additional natural number N, the set is represented by the set of all numbers smaller than N.

- $1 = \{ 0 \} = \{ \emptyset \}$
- $2 = \{ 0, 1 \} = \{ \emptyset, \{ \emptyset \} \}$
- $3 = \{0, 1, 2\} = \{ \emptyset, \{\emptyset\}, \{\emptyset, \{\emptyset\}\}\}$
- ...

Now we need to show that the axioms that define the meaning of the natural numbers are true when applied to this construction. For natural numbers, that means we need to show that the Peano axioms are true.

1. *Initial Value Rule:* The first Peano axiom that we'll look at is the *initial value rule,* which says that 0 is a natural number. In our construction of natural numbers, we've got a 0 that is a natural number. Given our 0, the initial value rule is satisfied.

2. *Successor, Uniqueness, and Predecessor Rules:* Peano arithmetic says that every number has exactly one unique successor. It should be obvious that in our construction, there's exactly one successor for each number. For any number N, the successor is created as the set of values from *0* to N. There's only one way to create a successor here, and it's clearly unique. If the successor rule is true, then the predecessor rule will *also* be true, as long as there is no predecessor for 0.

Since 0 is represented in our model as ∅, how could any set be its predecessor? The successor to any number N contains N as an element: the representation of 0 contains nothing, so it can't be the successor to anything.

3. *Equality Rules*: We'll use set equality. Two sets are equal if and only if they have the same elements. Set equality is reflexive, symmetric, and transitive, and that means that number equality in our set-based model will be the same.

4. *Induction:* The axiom of infinity is specifically designed to make inductive proofs work. It's a direct restatement of the induction rule in a stronger form. That means inductive proofs on the set-theoretic model of the natural numbers will work.

We've just built *a model* of the natural numbers using sets, and we showed that we can easily prove that the Peano axioms are true and valid for that construction. By doing that, we now know that our definition of the natural numbers is consistent and that it's possible to create objects that satisfy it. The fact that the set-based model is both consistent and satisfies the axioms of Peano arithmetic means that, using that model, any proof about the natural numbers can be reduced to a proof in terms of the axioms of ZFC. We don't need to re-create any foundations the way we did when we were setting up ZFC: we did that once, and now we'll just keep on using it.

Models from Models: From Naturals to Integers and Beyond!

I said that sets were like LEGOs, and I really meant it. To me, the way that you build structures using sets and LEGOs is very similar. When you're building an interesting project like a fancy house with LEGOs, you don't usually start by thinking about how you're going to build it in terms of individual bricks. You start by dividing your project into components. You build walls, roof and support beams, and then you'd put those together to make your house. To build the walls, you might divide a wall into sections: a base, a window, and side sections around the windows. In math, when we're building a model with sets, we often do the same thing: we build simple components from sets and build more elaborate models out of those components.

We built a nice model of the natural numbers using nothing but sets. In fact, we started with nothing but the empty set and then used the axioms of ZFC to allow us to build the set of natural numbers.

Now we'd like to move on and build more numbers. We do this exactly the same way that we did way back in 2, *Integers*, on page 9, but we're doing it using our set-based naturals.

What we're going to do is take the natural numbers that we've defined using sets. We're going to keep the same exact set of values that we did before, but we're going to assign a different meaning to them.

For every *even* number *N*, we're going to say that it represents a *positive* integer equal to *N/2*; for every odd integer, we're going to say that it represents a *negative* integer equal to *(N+1)/2*.

Of course, for the sake of formality, we do need to define just what it means to be even.

$$\forall\, n \in N:\ \text{Even}\,(n) \lneqq (\exists\, x \in N: 2 \times x = n)$$

$$\forall\, n \in N:\ \text{Odd}\,(n) \lneqq\ \neg\ \text{Even}\,(n)$$

So:

- The natural number 0 will represent the integer 0; 0 is even, because 0 * 0 = 0; so 0 is represented by the empty set.
- The natural number 1 will represent the integer –1. 1 is odd, so it represents –(1+1)/2 = –1. Therefore –1 is represented by the set *{ ∅ }*.
- 2 represents the integer +1, which means that +1 is represented by the set *{ ∅, { ∅ }}*.
- 3 is –2, so –2 is represented by *{ ∅, { ∅ }, { ∅, { ∅ }}*.
- And so on.

To be a valid model of the integers, we need to show that the axioms for integers hold for this model. It's pretty obvious that most of them do, because we verified them for the natural numbers. The one new thing that we've added is the idea of an additive inverse. So we need to show that the additive inverse axiom works for this model of the integers. The additive inverse axiom says that every integer has an additive inverse. By the definition of addition, we know that if both N and $-N$ exist, then $N + -N = 0$. What we have to show is that for every N greater than or equal to 0, our model does have a $-N$, and for every N less than or equal to 0, our model has a $+N$.

As usual for proofs about number theory, we'll use induction.

1. Base case: 0 is its own additive inverse, so the additive inverse of 0 exists.

2. Induction: for any integer n, if the additive inverse of n exists, then the additive inverse of $n + 1$ must exist.

3. We can show that our inductive case is true by saying that for any number n, it is represented by the natural number $2n$. Knowing that $2n$ represents n, then by the definition of our model of the integers, $2n+2$ represents $n+1$, and $2n+1$ represents the additive inverse of $n + 1$. We've shown that the additive inverse of $n+1$ exists by showing exactly how it's represented.

That little proof demonstrates one of the beauties of using sets as a LEGO model. We started off with nothing but the empty set. We used that to construct the natural numbers. Then we used the natural numbers to construct the integers. If we hadn't taken that step of building the natural numbers and then building the integers using the natural numbers, proving that there's an additive inverse would be a lot more painful. It definitely could be done, but it would be a lot harder to both read and write.

Similarly, we can keep going and build up more numbers. Back in 3, *Real Numbers*, on page 15, when we defined the rational numbers in terms of pairs of integers and real numbers as Dedekind cuts of rationals, we were really using set-based models of those kinds of numbers built on top of the very first set-based model of natural numbers. We're just being explicit about the fact that we've got a set-based model of the most primitive kind of number, and then we use the bricks of set-based constructions to build on that first model to build better and better bricks.

That's how we build math from sets. We just build up the blocks from simple pieces and then use those blocks to build better pieces until we get to what we want; whether it's numbers, shapes, topologies, or computers, it's all the same process of LEGO-building using sets.

There's one point about this that can't be reinforced enough. Sets are pretty much the ultimate mathematical building blocks. In this chapter, we built models of the natural numbers and the integers. We didn't build *the natural numbers*; we build *a model* of the natural numbers. The model of an object is *not* the object being modeled.

Think about LEGOs. You can build a great model of a car from LEGOs. But a car isn't made from LEGOs. If you're building a model of a car, what a car is really made of doesn't really matter. What you want to do is to build something that looks like a car, that acts like a car. Even if you build a giant LEGO model that people can ride in and it in fact really is a car, that doesn't mean that cars are made of LEGOs!

This is a point about building models with sets that constantly throws people off. We built a beautiful model of the natural numbers here. But the objects in our model that represent numbers are still sets. You can take the intersection of the *models* of the numbers 7 and 9 in our construction. But you can't take the intersection of the *numbers* 7 and 9, because the numbers aren't the objects in the model. When you're working with the objects in the model, you have to work with them entirely in terms of operations *in the model*. In the model, you have to stay in the model or else the results won't be meaningful.

Whenever you build anything in set theory, you have to remember that. You build models of object with sets, but the objects that you modeled aren't the models. You need to be careful to stay inside the model if you want valid results. To use the LEGO metaphor one more time: you can build a LEGO model of a house, using clear bricks for the windows. But you can remove a model window from the model house, break it apart into the individual bricks, and then reassemble it. You can't do that with a real window.

19

Transfinite Numbers: Counting and Ordering Infinite Sets

One of the deepest questions you can ask is the very first one that set theory was used to answer: Can you have something infinitely large that is *larger than* something else that is infinitely large? Once you know the answer to that question and recognize that there are degrees of infinity, that opens up a new question: How can you talk about infinity using numbers? How does arithmetic work with infinity? What do infinitely large numbers mean? We'll answer those questions in this section by looking at numbers through the lens of sets and at Cantor's definition of cardinal and ordinal numbers in terms of sets. This will lead us to looking at a new kind of number: Cantor's *transfinite numbers*.

Introducing the Transfinite Cardinals

Cardinality is a measure of the size of a set. For finite sets, that's a remarkably easy concept: count up the number of elements in the set, and that's its cardinality. When Cantor was first looking at the ideas of numbers and set theory, the first notion of the meaning of numbers that became clear was cardinality. If we have a set, one of the most obvious questions we can ask about it is "How big is it?" The measure of the size of a set is the *number* of elements in that set, which we call the set's *cardinality*.

Once we have a notion of the size of a set, we can start asking questions about the *relative* size of sets: "Which set is bigger—this one or that one?" That question becomes even more interesting when we get to infinitely large sets. If I have two infinitely large sets, is one bigger than the other? If there are different sizes of infinitely large sets, how many degrees of infinity are there?

We've already seen an example of how to compare the cardinality of different infinite sets in Section 16.2, *Cantor's Diagonalization*, on page 131. The idea of measuring relative cardinality is based on the use of one-to-one functions between sets. If I have two sets S and T and if there is a perfect one-to-one mapping from S to T, then S and T have the same cardinality.

It seems like a simple notion, but it leads to some very strange results. For example, the set of even natural numbers has the same cardinality as the set of natural numbers: $f(x) = 2*x$ is a total, one-to-one function from the set of naturals to the set of even naturals. So the set of even naturals and the set of all naturals have the same size, even though the set of evens is a proper subset of the set of natural numbers.

When we look at most sets, we can classify them into one of three cardinality classes: finite sets, whose cardinality is smaller than the cardinality of the natural numbers; countable sets, which have the same cardinality as the set of natural numbers; and uncountable sets, which have a cardinality greater than the cardinality of the natural numbers.

Set theorists, starting with Cantor, created a new kind of number just for describing the relative cardinalities of different sets. Before set theory, people thought that for talking about sizes, there were finite numbers, and there was infinity. But set theory showed that that's not enough: there are different infinities.

To describe the cardinalities of sets, including infinite sets, you need a system of numbers that includes something more than just the natural numbers: Cantor proposed an extended version of the natural numbers that he called *cardinal numbers*. The cardinal numbers consist of the natural numbers plus a new family of numbers called *transfinite* cardinal numbers. The transfinite cardinal numbers specify the cardinality of infinite sets. The first transfinite number is written $\aleph_0$ (pronounced "aleph-null"), and it's the size of the set of natural numbers.

When you start looking at the transfinite numbers and trying to work out what they mean about infinity, you get to some fascinating results. The same basic idea that we used in Cantor's diagonalization can be used to prove that for any non-empty set *S*, the set of all subsets of *S* (also called the powerset of *S*) is strictly larger than *S*. That means that given the smallest infinite set, $aleph_0$, you can prove that there's got to be another infinite set larger than it; we'll call that $aleph_1$. Then given $aleph_1$, there's another infinite set $aleph_2$ which is bigger than $aleph_1$, and so on. There's an infinite cascade of ever-larger infinities!

The Continuum Hypothesis

Cantor proposed that the first infinite set larger than $\aleph_0$ was of size of the powerset of $\aleph_0$, which is the size of the set of reals. That proposition is known as the *continuum hypothesis*. If the continuum hypothesis is true, then

$$\aleph_1 = 2^{\aleph_0}$$

The continuum hypothesis turns out to be a really sticky problem. In the model of numbers constructed from set theory (and thus, in all set-theoretic mathematics!), it's *neither true nor false*. That is, you can choose to treat it as true, and all of ZFC mathematics will be fine, because you'll never be able to prove a contradiction. But you *also* can take it as being *false*, and still you won't find any contradictions in ZFC.

Looking at the continuum hypothesis, you might think that it's a problem like Russell's paradox in naive set theory. After all, it looks like we've got something that's somehow both true and false!

But it's *not* really a problem. In Russell's paradox, we had a question that had two possible answers, and they were both provably false. In the continuum hypothesis, we have a question with two possible answers, and *neither answer* is provable. There's no contradiction here: we can't create any proofs that are inconsistent. We've just discovered one of the limits of ZFC set theory: the continuum hypothesis isn't provable either way. There are perfectly valid systems of transfinite numbers where it's true and where it's false: we can choose either of the two equally valid systems by adding our choice on the continuum as an axiom.

That's just weird. But that's the nature of math. As I've said before, the history of math is full of disappointments. When the continuum problem was proposed, lots of people believed that it was true; lots of people believed it was false. But no one would have dreamed that it could be independent, that it could be neither provably true nor provably false. Even as we define math from the simplest, clearest basis, we can't get away from the fact that nothing works quite the way we expect it to.

Where in Infinity?

Now we've got sets, and we can talk about the size of those sets using cardinal numbers. But even if we have a set where we know its size and we know that its elements are ordered by some kind of relation, we can't talk about *where* in the set a particular value lives. A cardinal number describes the number of elements in a set. It's a way of specifying a quantity.

In English, it's easy to show why you can't use ordinals in the place of cardinals, but it's much harder to do the other way around. If you speak English and you try to say "I have seventh apples," it's obviously wrong. Going the other way, you can say "I want apple 3," and it seems like you're using a cardinal to specify an ordinal position. In English, you can get away with that. In math, you can't. You need a *different* object, with a different meaning, to refer to a position than you do to refer to a measure of quantity.

We need to define ordinal numbers. We can use the same representation to build a model of the ordinals. It's OK that we're using the same objects under the hood for the ordinals and the cardinals: remember that you can only talk about the modeled object *in the model*. So we can't take a cardinal from the cardinal model and drop it into the ordinal model and expect it to make sense.

Now that we've got the ordinal numbers, what happens when we try to use them for sets with infinite cardinality? How can we describe ordinal positions within a set whose size is $\aleph_0$? To talk about the position of elements inside that, we need some way to represent the first position of an element *after* all of the finite ordinal positions.

Just like we needed to define transfinite cardinals, we'll need to define a new kind of ordinal number called a *transfinite ordinal* number. We use the symbol ω ("omega") for the first transfinite ordinal.

Transfinites are one of the places where the behavior of the ordinals becomes very different from the cardinals. If you add one element to a set whose size is $\aleph_0$, the size of the set is *still* $\aleph_0$. But if you look at position ω and then look at the position after it, position $\omega + 1$ does come after position ω, which means that $\omega + 1$ is *greater than* $\omega + 1$.

Remember, we're not talking about size, we're talking about position; and even when we get to the transfinite realm, there can be an object *next to* an object in position ω. Since it's in a distinct position, it needs a distinct transfinite ordinal. When we talk about ordinals, there are three kinds of ordinal numbers:

Initial Ordinal The initial ordinal is 0, which is the position of the initial element of a well-ordered set.

Successor Ordinals Successor ordinals are ordinals that we can define as the next ordinal after (aka the successor to) some other ordinal. All finite numbered positions are successor ordinals.

Limit Ordinals Limit ordinals are ordinal numbers like ω that you can't get to by taking any number of successors.

ω is a limit ordinal. It's the limit of the finite ordinals: as the first non-finite ordinal, every finite ordinal comes before it, but there is no way of specifying just what ordinal it's the successor to. (There is no subtraction operation in ordinal arithmetic, so ω-1 is undefined.) Limit ordinals are important, because they're what gives us the ability to make the connection to positions in infinite sets. A successor ordinal can tell us any position within a finite set, but it's no good once we get to infinite sets. And as we saw with the cardinals, there's no limit to how large sets can get, because there's an infinite number of transfinite cardinals with corresponding sets.

So how do we use transfinite ordinals to talk about position in sets? In general, it's part of a proof using *transfinite induction*. So while we can't necessarily specifically identify element ω of a set with transfinite cardinality, we can talk about the ωth element.

The way that we do that is by isomorphism. In mathematical terms, an *isomorphism* is a strict one-to-one mapping between two different sets. Every well-ordered set is isomorphic to the set-form of an ordinal. A set with N elements is isomorphic to the ordinal $N+1$.

We can talk about the ωth element of an infinite set by talking in terms of the well-ordering and the isomorphism. We know that sets of this size exist; so if the set exists, the ordinal must as well.

Now we've got our ordinal numbers. And you can see pretty easily that they're different from cardinal numbers. The cardinal $\aleph_0$ is the cardinality of the set representation of ω. And of $\omega + 1$. And $\omega + 2$. And so on. So in ordinals, ω is different from $\omega + 1$. But in cardinals, the sizes of ω and $\omega + 1$ are the same, because $\aleph_0$ is the same as $\aleph_0 + 1$.

They're different, because they mean different things. It's not so obvious when you're looking at finite values. But once you hit infinity, they're completely different and follow completely different rules.

20

Group Theory: Finding Symmetries with Sets

Every day, most of us get up in the morning, go into the bathroom, and brush our teeth in front of a mirror. When we do that, we're confronted with a profound mathematical concept that we rarely think about: symmetry.

Mirror reflection is the kind of symmetry that's most familiar to us because we experience it every day. There are many other kinds of symmetry: symmetry is a very general concept that comes up everywhere from math to art to music to physics. By understanding the mathematical meaning of symmetry, we can see how all of the different kinds of symmetries are all really different expressions of the same fundamental concept.

To understand what symmetry really is, we'll use the LEGO-like nature of sets to build an algebraic structure called a *group* to explore *group theory*. Groups capture what it means to be *symmetric* in precise, formal mathematical terms.

Puzzling Symmetry

Let's play a game that I'm calling crypto-addition. I've taken the letters *a* through *k* and I've assigned each one to one of the numbers between -5 and +5, and I want you to figure out which letter represents which number. The only clues come from *A crypto-addition table,* where I've shown you the sums of different letters whenever that sum is between -5 and +5.

	a	b	c	d	e	f	g	h	i	j	k
a:			b	a	h		f		d	k	
b:			f	b	d		k	a	c	h	
c:	b	f	g	c	i	k	j	d		e	h
d:	a	b	c	d	e	f	g	h	i	j	k
e:	h	d		e		c		j			g
f:			k	f	c		h	b	g	d	a
g:	f	k	j	e		h	e	c		i	d
h:		a	d	h	j	b	c	k	e	g	f
i:	d	c		i		g		e			j
j:	k	1	e	j		d	i	g			c
k:			h	k	g	a	d	f	j	c	b

Table 3—A crypto-addition table

The letters *a* through *k* represent the numbers between -5 and +5. Entries in the table represent the sums of the row and column labels.

Using the crypto-addition table, what can you figure out about which numbers are represented by which letters?

You can easily figure out what zero is: just look at the first row. This shows you what you get for each symbol when you add the number represented by *a* with every other number. When you add *a* to *d*, you get *a*, so *d* is zero, because zero is the only number where $x + 0 = x$.

Once you know what zero is, you can figure out pairs of positive and negative numbers, because they'll add up to zero. So we don't know what number *a* is, but we do know that $a + i = 0$, and that therefore $a = -i$.

What else can you figure out? With enough experimentation, you can find an ordering: if you start with *a* and add *c* to it, you get *b*. If you keep adding *c* to it again, you get *f*, *k*, *h*, *d*, *c*, *g*, *j*, *e*, and finally *i*. Or you could start with *i* and then repeatedly add *h* to it, which will give you the opposite ordering. Only *h* and *c* can step through the entire sequence of numbers, so we know that they're *-1* and *1*, but we don't know which is which.

Knowing the ordering, we can figure out most of the rest: we know that *c* and *h* are *-1* and *1*; *g* and *k* are -2 and 2; *f* and

j are -3 and 3; and b and e are -4 and 4. We already knew that a and i were -5 and 5.

Which symbol represents +1, c or h?

You can't tell. a might be -5, in which case c is +1, or a might be +5, in which case c would be -1. Since our only clue is addition, there's absolutely no way that we can tell which numbers are positive and which are negative. We know that i and j are the same sign, but we don't know which sign that is!

The reason that we can't tell which sign is which is because addition of integers is *symmetric*. You can change the signs of the numbers, but according to everything you can do with addition, the change is invisible.

Using my letter representation, I can write an equation, $a + c = b$, knowing that $a = 5$, $b = 4$, and $c = -1$: $5 + -1 = 4$. Then I can switch all of the signs in my representation, and the equation still works: $-5 + 1 = -4$. In fact, any equation which relies only on addition can't tell the difference when the signs are switched. When you looked at the crypto-addition puzzle, the symmetry of addition meant that nothing you could do could tell you which numbers were positive and which were negative.

That's a simple example of what symmetry means. Symmetry is an *immunity to transformation*. If something is symmetric, that means that there is something you can do to it, some transformation you can apply to it, and after the transformation, you won't be able to tell that any transformation was applied.

That basic idea of a set of values with one operation is the heart of what's called *group theory*. Group theory is all about symmetry, and it's built around the fact that every kind of symmetry can be described in terms of a group.

A group is exactly what we were looking at in the example before. It's a set of values with one closed operation. To be formal, a group consists of a tuple $(S, +)$, where S is a set of values, and $+$ is a binary operation with the following properties.

Closed For any two values, $a, b \in S$, $a + b \in S$.

Associativity For any three values, $a, b, c \in S$, $a + (b + c) = (a + b) + c$.

Identity There is a value $0 \in S$, where for any value $s \in S$, $0 + s = s + 0 = s$.

Inverse For every value a in S, there is another value b in S where $a + b = b + a = 0$. b is called the *inverse element* of a.

If the operation satisfies those rules, then you have a group. When something is a group, there is a transformation associated with the group operator that is undetectable within the structure of the group. Exactly what that transformation is depends on the specific values and operation that's associated with the group. For convenience, most of the time when we're talking about a group, we'll write the group operation using a plus sign, but it doesn't have to be addition: the group operation can be anything that satisfies the constraints we listed. We'll see some examples in a bit that use different operations to form a group.

When we talk about symmetry in groups, we say that the operation of a group produces an effect that is invisible *within the group*. The "within the group" part is important: when you apply an operation other than the group operation, you can detect the change. For example, with our crypto-arithmetic puzzle, if you could do multiplication, then you could tell c apart from h: c times b is e, and c times e is b, which would tell you that c must be -1. Multiplication isn't part of the group, so when you use it, you're no longer symmetric.

Think back to the intuitive idea of mirror symmetry: the integer addition group is the mathematical description of a mirror! What mirror symmetry means is that if you draw a line through an image and swap what's on the left-hand side of it with what's on the right-hand side of it, the mirror-symmetric image will be indistinguishable from the original image. That's exactly the notion that we've captured with the group formed by real numbers with addition. The construction of addition-based groups of numbers captures the fundamental notion of mirror symmetry: it defines a central division (0), and it shows how swapping the objects on opposite sides of that division has no discernible effect. The only catch is that we're talking about integers, not images. Later on, we'll look at how to take the symmetry defined by a group and apply it to other kinds of things, like applying the mirror symmetry of integer addition to a picture.

Different Kinds of Symmetry

There are more kinds of symmetry than just mirrors. For example, look at the folloiwng figure, which shows multiple symmetries on a hexagon. A hexagon has two different kinds of mirror symmetries. But going beyond the mirror symmetry, it also has *rotational* symmetry: if you rotate a hexagon by 60 degrees, it's indistinguishable from the hexagon before rotation.

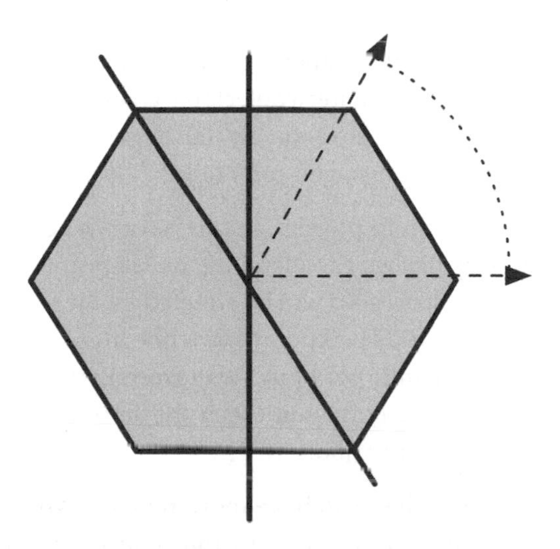

Figure 12—Multiple symmetries—hexagon: *Hexagons have multiple symmetries—two kinds of mirror symmetry and a rotational symmetry.*

All sorts of transformations can be described by groups as long as there's some way in which the transformation can be performed without producing a visible effect. For example, all of the following are possible symmetric transformations that can be described by groups.

Scale Symmetry of scale means that you can change the size of something without altering it. To understand this, think of geometry, where you're interested in the fundamental properties of a shape—the number of sides, the angles between them, the relative sizes of the sides. If you don't have any way of measuring size on an absolute basis, then an equilateral triangle with sides 3 inches long and an equilateral triangle with sides 1 inch long can't be distinguished. You can change the scale of things without creating a detectable difference.

Translation Translational symmetry means you can move an object without detecting any change. If you have a square grid like graph paper drawn on an infinite canvas, you can move it the distance between adjacent lines, and there will be no way to tell that you changed anything.

Rotation Rotational symmetry means you can rotate something without creating a detectable change. For example, if you rotate a hexagon by 60 degrees without any external markings, you can't tell that it's rotated.

Lorentz Symmetry In physics, if you have a laboratory in a spaceship that isn't accelerating, no lab result conducted inside the spaceship will be affected by the speed of the ship. If you did an experiment while the ship was traveling at one mile per hour away from Earth, the results would be exactly the same as if the ship were traveling at 1000 miles per hour.

A single set of values can have more than one kind of symmetry by forming it into groups that pair it with different operations. For example, in Figure 13, *Multiple symmetries—a tiling pattern,* on page 173, you can see at least four basic symmetries: mirror symmetry, rotation symmetry, translation symmetry, and color-shift symmetry.

Groups start to explain in a mathematical way what symmetry *is*. The way we've looked at it so far has one big problem: it's limited to things that we can manipulate algebraically. We defined mirror symmetry using numbers and addition, but when we think about mirror symmetry, we're not thinking about numbers! We're thinking about pictures and reflections, about what things that are symmetric *look like*. How can we take this basic idea of group symmetry and

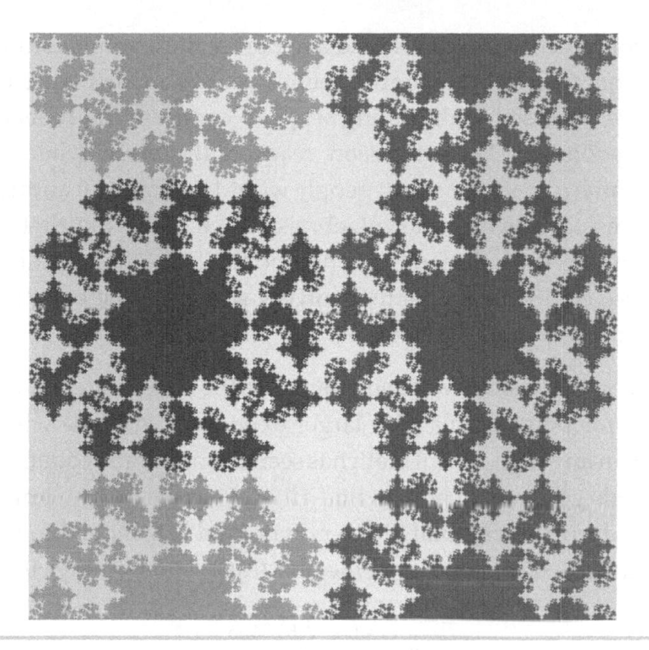

Figure 13—Multiple symmetries—a tiling pattern: *This tiling pattern has many symmetries: mirror, rotation, translation, and color-shift.*

expand it so that it makes sense for talking about real-world objects?

What we can do is introduce something called a *group action*. A group action allows you to take the symmetry of a group and apply it to other things. To understand the group action and how to use groups to describe symmetry in general, the easiest approach is to look at something called *permutation groups*.

Before we go into the formal math of the permutation groups and how they lead to group actions, we're going to take a brief diversion into history to see where group theory came from.

Stepping into History

Before group theory was invented, the notion of symmetry was first captured mathematically in what we now call the permutation groups. The study of permutation groups was the reason group theory was originally created!

Group theory was developed as a part of the algebraic study of equations. During the nineteenth century, mathematicians were obsessed with finding ways to compute the roots of polynomials. This obsession reached the point where it became a spectator sport: people would assemble in auditoriums to watch master algebraists race to be the first to compute the solutions to a polynomial! The holy grail to these competitive equation-solvers was a simple equation that could yield the solutions to a polynomial.

What they wanted was something like the quadratic equation but for polynomials with larger exponents. Anyone who's taken any high-school math has seen the quadratic equation: it tells you exactly how to find all of the roots of any simple quadratic polynomial. If the polynomial is arranged in the order $ax^2 + bx + c = 0$, then the quadratic equation says this:

$$x = \frac{-b \pm \sqrt{b^2 - 4ac}}{2a}$$

Just fill in the specific values of a, b and c, and do the arithmetic, and you've got the roots of the equation! There are no tricks that you need to know how to do, no clever manipulations of the polynomial, no factoring or rewriting. We're used to it, so it doesn't seem to be a big deal, but being able to just extract the solution to a polynomial with a simple mechanical process is an amazing thing!

The quadratic solution has been known for a very long time—there are records dating back to the Babylonians that contain forms of the quadratic equation. Even starting from scratch, deriving the quadratic equation is easy enough that many high-school math classes go through the derivation. Since it's so easy for the quadratic, you might think that doing the same thing for polynomials with higher powers might not be too difficult. It turns out that it's *very* difficult. The quadratic equation was known hundreds of years before the common era, but it took until the middle 1500s to find a solution to the cubic (third power) and quartic (fourth power) equations. After the quartic was discovered in 1549, there was no progress in finding root equations for hundreds of years.

Finally, in the nineteenth century, two men named Niels Henrik Abel (1802–1829) and Évariste Galois (1811–1832), both very young and very unlucky mathematicians, simultaneously proved that there was no general solution for quintic equations (fifth power). Galois did it by recognizing that there are fundamental symmetries in the solutions of polynomials. By identifying what those symmetry properties were, he showed that you couldn't derive a single equation to solve all of the quintic equations. He showed that there was no possible way to get a general solution because of the properties of permutation groups for those equations.

Abel and Galois were both amazing young men, and both died tragically young.

As a young student in Norway, Abel derived what he believed to be the solution to the quintic polynomials. He later went on to discover the error in his solution and to broaden that into a description of the symmetry properties of polynomial equations that showed that there was no general solution for polynomials with degrees higher than four. While traveling to Paris to present his work to the French academy of mathematics, he became infected with tuberculosis, the disease that eventually killed him. On a trip home for his own wedding, weakened by his tuberculosis, he became ill with pneumonia and finally died. While he was home, dying, he finally received an appointment to a professorship of mathematics at a university in Berlin, but he never learned of it and never received the recognition he deserved for his work.

The story of Galois is even sadder. Galois was a mathematical prodigy. He started publishing his own original mathematical work on the subject of continued fractions when he was just sixteen years old! He submitted his first paper on the symmetric properties of polynomials just one year later, at the ripe old age of seventeen. Over the next three years, he wrote three papers that defined the entire basis of what became group theory. Just one year after that, he died in a duel. The exact circumstances aren't known for certain, but from letters he sent days before his death, it appears that one of the greatest mathematical minds of the nineteenth century died at the age of twenty-one in a duel over a failed love affair.

The Roots of Symmetry

Galois and Abel independently discovered the basic idea of symmetry. They were both coming at the problem from the algebra of polynomials, but what they each realized was that underlying the solution of polynomials was a fundamental problem of symmetry. The way that they understood symmetry was in terms of *permutation groups*.

A permutation group is the most fundamental structure of symmetry. As we'll see, permutation groups are the master groups of symmetry: every kind of symmetry is encoded in the structure of the permutation group.

Formally, a permutation group is a simple idea: it's a structure describing all of the possible permutations, or all of the possible ways of rearranging the elements of a set. Given a set of objects, O, a permutation is a one-to-one mapping from O to itself, which defines a way of rearranging the elements of the set. For example, given the set of numbers $\{1, 2, 3\}$, a permutation of them is $\{1 \to 2, 2 \to 3, 3 \to 1\}$. A *permutation group* is a collection of permutations over a set, with the composition of permutations as the group operator.

For example, if you look at the set $\{1, 2, 3\}$ again, the elements of the largest permutation group are $\{ \{ 1 \to 1, 2 \to 2, 3 \to 3 \}, \{ 1 \to 1, 2 \to 3, 3 \to 2 \}, \{ 1 \to 2, 2 \to 1, 3 \to 3 \}, \{ 1 \to 2, 2 \to 3, 3 \to 1 \}, \{ 1 \to 3, 2 \to 1, 3 \to 2 \}, \{ 1 \to 3, 2 \to 2, 3 \to 1 \} \}$.

To see the group operation, let's take two values from the set. Let $f = \{1 \to 2, 2 \to 3, 3 \to 1\}$ and let $g = \{1 \to 3, 2 \to 2, 3 \to 1\}$. Then the group operation of function composition will generate the result: $f * g = \{1 \to 2, 2 \to 1, 3 \to 3\}$.

To be a group, the operation needs to have an identity value. With permutation groups, the identity is obvious: it's the null permutation: $1_O = \{1 \to 1, 2 \to 2, 3 \to 3\}$. Similarly, the group operation needs inverses, and the way to get them is obvious: just reverse the direction of the arrows: $\{ 1 \to 3, 2 \to 1, 3 \to 2 \}^{-1} = \{ 3 \to 1, 1 \to 2, 2 \to 3 \}$.

When you take the set of permutations over a collection of N values, the result is the largest possible permutation group over those values. It doesn't matter what the values are: every collection of N values effectively has the same permutation group.

That canonical group is called the *symmetric group* of size N, or S_N. The symmetric group is a fundamental mathematical object: every finite group is a subgroup of a finite symmetric group, which in turn means that every possible symmetry of every possible group is embedded in the structure of the corresponding symmetric group. Every kind of symmetry that you can imagine, and every kind of symmetry that you *can't* imagine, are all tied together in the symmetry groups. What the symmetry group tells us is that all symmetric groups are really, under the hood, different reflections of *the same thing*. If we can define a kind of symmetry on one set of values with a group operation, then by the fundamental relationship illustrated by the symmetric group, we can apply it to *any* group, as long as we do the mapping in a way that preserves the fundamental structure of the symmetric group.

To see how that works, we need to define a subgroup. If you have a group $(G,+)$, then a subgroup of it is a group $(H,+)$ where H is a subset of G. In English, a subgroup is a subset of the values of a group that uses the same group operator and that satisfies the required properties of a group.

For example, given the group made from the set of real numbers combined with addition as the group operation, then the set of integers is one of its subgroups. We can show that the integers with addition are a group because it's got the necessary properties: any time you add any two integers, the result is an integer, which means that its arithmetic is closed over addition, the arithmetic inverse of an integer is an integer, and so on. You can work through the other two properties required of a group, and it will satisfy them all.

We're finally ready to get to the general symmetry of groups. As we said before, groups define a particular kind of symmetry as immunity to transformation. Defining that precisely for a specific collection of values, figuring out the group operator, and showing that it all works properly is a lot of work. We don't want to have to define groups and group operators for every set of values that we see as symmetric. What we'd like to do is capture the fundamental idea of a kind of symmetry using the simplest group that really exhibits that kind of symmetry and then able to use that group as the definition of that kind of symmetry. To do that, we need to be able to describe what it means to apply the symmetry defined by a group to some set of values.

We call the transformation of a set produced by applying a symmetric transformation defined by a group G as the *group action* of group G.

Suppose we want to apply group G as a symmetric transformation on a set A. What we can do is take the set A and define the symmetric group over A, S_A. Then we can define a specific type of strict mapping called a *homomorphism* from the group G to S_A. That homomorphism is the *action* of G on the set A. In our definition (following) of the group action, the constraint basically says that the action preserves the structure of the symmetry group. Here it is in formal terms:

If $(G,+)$ is a group and A is a set, then the *group action* of G on A is a function f such that

1. $\forall g \in G : \forall a \in A : f(g + h, a) = f(g, f(h, a))$
2. $\forall a \in A : f(1_G, a) = a$

All of which says that if you've got a group defining a symmetry and a set you want to apply a symmetric transformation to, then there's a way of mapping from the elements of the group to the elements of the set, and you can perform the symmetric group operation through that map. The group action is an application of the group operation through that mapping.

Let's see how that works in terms of an image. In Figure 14, *Mapping a group to a set*, on page 179 you can see a mirror-symmetric image. Mirror-symmetry is described by the group of integers with addition. The way that we can show the symmetry of the image with the group action is to divide the image into rows. On each row, we take the blocks and assign a mapping between the blocks and the integers. Clearly, the result is a *subgroup* of the integers.

To see the symmetry in terms of the group action, we can divide it into columns and map those columns onto integers. The symmetry of the image is exactly the symmetry of the integers in the mapping. To illustrate that in the image, I've

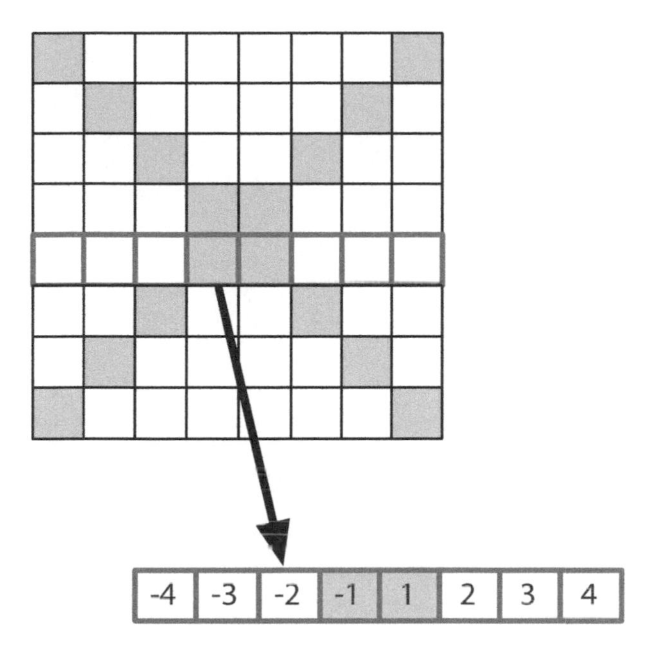

Figure 14—Mapping a group to a set: *The relationship between the symmetry of the integers and the mirror symmetry of a simple bitmap image is illustrated here by showing the mapping that underlies the group action.*

pulled out one row and written the mappings onto the blocks.

By exploiting the deep relation between *all* symmetric structures, we can apply the symmetry of any group to any other group of values using the group action. It's amazing to realize, but every kind of symmetry we can imagine is really exactly the same thing. From mirrors to the theory of special relativity, it's all the same simple, beautiful thing.

Part VI

anical Math

ath should be obvious given what I do
r scientist, and I've spent my career
ular, I've worked as both a researcher
different companies, building software
software, like programming languages,
d development tools.

So what area of math am I talking about? *Computation.* Computation is the branch of math that studies what kinds of things machines can do. Long before anyone had actually built a real computer of any kind, mathematicians and logicians set the groundwork for what became the field of computer science. They designed theoretical machines, computing systems, and logics, and they studied what was possible using those theoretical constructs. In addition to being a theoretical model of the machines that we use, the theory of computing formed the foundation of the programming languages that are my obsession.

In this part of the book, we're going to take a look at the mechanical marvels that were designed by these people and see how we can use them to understand what machines can do and what they can tell us about not just machines, but about information in the world we live in.

We'll begin with a quick tour of different kinds of computing devices. We'll start with something so simple that it can't even count, but that

I and many other programmers around the world use every day. Then we'll make a jump to the Turing machine, which is one of the fundamental theoretical machines that is used to study the limits of computation. Then for a bit of fun, we'll peek at a variant of something called the P'' (pronounced "P prime-prime") machine, which was implemented as one of the most bizarrely delightful programming languages ever designed.

Finally, we'll end this survey by looking at λ calculus. λ calculus is another foundational theoretical model. It's a bit harder to understand than the machines because it's less concrete, but it's widely used in practice. I use it every day! At my job, I use a programming language called Scala, and Scala is just a fancy syntax for λ calculus.

21

Finite State Machines: Simplicity Goes Far

When mathematicians and computer scientists try to describe computation, we start with very simple machines and then gradually add capabilities, creating different kinds of machines that have different limits until we get to the most complex machines that can be built.

We're not going to go through all of the possible classes of computing devices—that would take a much longer book than this one! Instead, we'll look at the two types of machines that form the limits: the most limited simple machines and the most powerful complex ones.

We'll start looking at computing by looking at the simplest kind of computing machine that does anything useful. This type of machine is called a *finite state machine* (or FSM for short). In formal writing, it's also sometimes called a *finite state automaton*. A finite state machine's computational abilities are limited to scanning for simple patterns in a string of characters. When you see how trivial they are, it might be hard to imagine that they're useful for much. In fact, as we'll see, every computer you've ever used is really just a very complicated finite state machine.

The Simplest Machine

Finite state machines are very limited. They can't count, and they can't recognize deep or nested patterns. They really don't have much in the way of computational power. But they're still very useful. Every modern programming language has finite state machines in the form of regular expressions built in or provided by a library.

So let's look at how the machines work. A finite state machine really only does one thing: it looks at a string of characters and determines whether or not that string of characters fits some pattern. To do this, the machine has a small piece of state consisting of a single atomic value, called the *state* of the machine. When it's performing a computation, the FSM is allowed to look at exactly one character of input. It's not allowed to sneak a peak ahead, and it's not allowed to look at any previous characters to see how it got into its current state. It just iterates over the characters in order, looking at each character exactly once, and then answers either yes or no.

Let's get precise.

An FSM processes strings of symbols in a specific *alphabet*. For example, in most programming languages, you can define regular expressions that work on either ASCII characters or unicode codepoints.

The machine itself consists of these parts:

- A set of *states*, which we'll call S.
- One special state, i from S, that we call the *initial state*—whenever we try to process a string using a finite state machine, the machine starts in this state.
- A subset of S that we'll call f—these are the *final states* of the machine. If after processing all of the characters in an input string, the machine is in one of the states in this set, then the machine's answer for the string will be yes; otherwise it will answer no.
- Finally, there's t, which is the machine's *transition relation*—the transition relation defines how the machine behaves. It maps pairs of machine states and input symbols to target states. The way that it works is this: if there is a relation $(a, x) \rightarrow b$, that means that when the machine is in state a and it sees an input symbol x, it will switch to state b.

The machine starts to look at an input string in state *i*. For each symbol in the input string in sequence, it performs a single transition, consuming that input symbol. When it's consumed every symbol in the input string, if it's in a state that's part of the set *f*, then it *accepts* the string.

For example, we can create a machine that accepts strings that consist of any string containing at least one *a*, followed by at least one *b*.

- The alphabet for our machine is made up of the characters *a* and *b*.
- The machine has four states: *{ 0, A, B, Err }*. Of these, *0* is the initial state, and *b* is the only final state.
- The state relation is shown in the following table about the AB finite state machine.

FromState	Char	ToState
0	a	A
0	b	Err
A	a	A
A	b	B
B	a	Err
B	b	B
Err	a	Err
Err	b	Err

In general, we don't write out the table like that; we can just draw the machine. The machine we're talking about is shown in Figure 15, *The AB finite state machine*, on page 186. Each state gets drawn as an oval, the initial state is marked with an arrow, the final states are marked with either a double outline or a bold outline, and the transitions are drawn as labeled arrows.

Let's step through a couple of input strings to see how it works.

- Suppose the input string is *aaabb*.
  1. The machine starts in state 0.

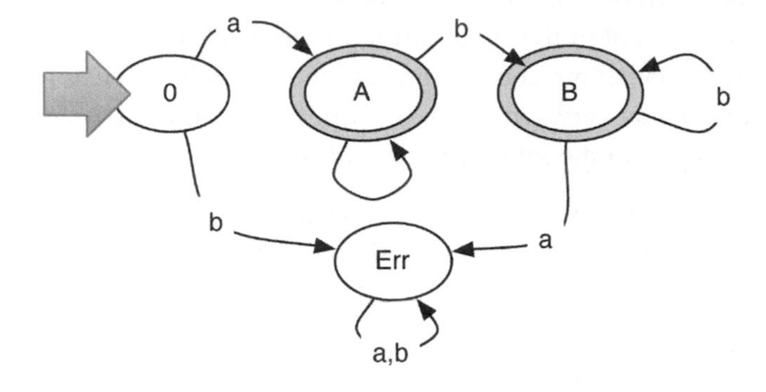

Figure 15—The AB finite state machine

2. It consumes the first character in the string, *a*, and it takes the transition from *0* to *A*.
3. The remaining input is *aabb*. It follows the transition for *a* from state *A* to *A*.
4. The remaining input is now *abb*, so now the machine processes the next *a*. It does exactly the same as the previous step, so it ends up staying in state *A*.
5. The remaining input is *bb*. It consumes a *b*, following the transition from *A* to *B*.
6. The remaining input is now just *b*. It consumes the last *b*, following the transition from *B* to *B*.
7. There's no input left. The machine is in state *B*, which is a final state, so the machine accepts the string.

- Suppose the input is *baab*.
  1. The machine starts in state 0.
  2. It consumes the first *b*, which moves it to state *Err*.
  3. The rest of the characters get consumed one at a time, but all of the transitions from *Err* in this machine return to *Err*: it's a dead end. Once it's consumed all of the characters, the machine will still be in *Err*, which isn't a final state, so it won't accept the string.

- Suppose the input is the empty string.
  1. The machine starts in state 0.
  2. Since there are no characters in the input, the machine never runs any transitions, so it ends in state *0*. Since state *0* isn't a final state, it doesn't accept the empty string.

When you look at this, it's really a trivial machine. It seems like there's very little that it can do. And yet, at least in theory, any computation that can be performed with a fixed, finite amount of state can be implemented with a machine this simple. Everything that I can do on the computer that I'm typing this on is doable using a finite state machine. It would have a huge number of states and an insanely complex state transition function, but it would be doable.

Finite State Machines Get Real

FSMs are great, but if you want to use something like an FSM in a real program, you probably don't want to try to figure out the whole transition relation and write it all out in code. When we really use a language processed by an FSM, the transition relation contains hundreds of transitions. It's just hard to write it out correctly. Fortunately, we don't have to. There's another way of writing an FSM: regular expressions. If you're a programmer, you're almost certainly already familiar with regular expressions: they're ubiquitous in real programs, and they're the way that we use FSMs.

Regular expressions aren't quite a way of writing the FSM. They're a syntax for writing down the language that an FSM should accept. Given a regular expression, you can translate it into many different possible FSMs, but they'll all end up doing the same thing. That's how programming languages do regular expressions: the programmer writes a regular expression and passes it to a compiler in a regular expression library, and the regular expression library translates regular expressions into FSMs, and then the regular expressions run the FSMs on input strings.

We're going to look at the simplest version of regular expressions. In most regular expression libraries, they have many more options for how to write things than what we're going to walk through. That's OK, because all of the additional features in the fancy regular expression syntaxes are just shorthands for the simple syntax that we'll use. The simple syntax is easier for us to talk about because we don't need to consider so many different options. In our syntax, a regular expression consists of the following:

Literal Characters A literal is just a character from the alphabet, such as *a*. A literal character exactly matches the character.

Concatenation If R and S are regular expressions, then RS is a regular expression. The language matched by RS is the concatenation of a string matched by R followed by a string matched by S.

Alternation Alternation describes a choice. If R and S are regular expressions, then $R|S$ is a regular expression, that matches anything matched by either R or S.

Repetition (aka Kleene closure) If R is a regular expression, then R^* is a regular expression. R^* matches any sequence of zero or more strings matched by R concatenated together.

As a shorthand, we can also write R^+, which means the same thing as RR^*, to match at least one repetition of something matched by R.

You can also use parentheses to group expressions in alternation, to make it easier to write choices between larger alternatives.

A few examples of regular expressions:

$(a|b|c|d)^*$ Any string of any length made from the characters *a*, *b*, *c*, and *d*, so it matches things like *abcd*, *aaaa*, *ab*, *dabcbad*, and so on

$(a|b)^*(c|d)^*$ A string of any length made of *a*'s and *b*'s, followed by a string of any length of *c*'s and *d*'s so it matches things like *ababbbacdcdcccc*, *ab*, *b*, *c*, *cdcdddcc*, and *ddcc*

$(ab)^+(c|d)^*$ A string of any number of repetitions of the sequence *ab* followed by any number of *c*'s and *d*'s—this matches *ababababcccd*, *abcc*, *ab*, and so forth.

$a^+b^*(cd)^*(e|f)$ Strings consisting of at least one *a*, followed any number of *b*'s (including zero), followed by any number of repetitions of *cd*, followed by either a single *e* or a single *f*.

When you look at finite state machines and regular expressions and you consider how they actually work, they seem to be so trivial that they shouldn't be useful for much. But that's not the case at all. Libraries that let programmers use regular expressions are one of the first things that any language designer adds to a new programming language, because no one would use a language without it.

Regular expressions make programmers' lives much easier. In real code, we constantly need to decompose strings based on some pattern. Regular expressions make that really easy to do. For example, I used to work on a program that described how to build complex software systems consisting of many components. One of the basic pieces of that system was called a target. A target was a string that specified a component in the system formatted as a sequence of directory names followed by the name of a file containing build directives, followed by the name of a specific directive in that file, like "code/editor/buffer/BUILD:interface." One of the things that I needed to do all the time was take a target and split it into the three pieces: the directory path, the filename, and the target name. So I put together a regular expression: "(.*)+/([A-Za-z]+):([A-Za-z]+)." The part of the string that matched the ".*" was the directory path; the part in the second parenthesis was the filename, and the part in the last parenthesis was the target name. That's typical of how we use regular expressions, and it's a facility that no programmer wants to live without.

Bridging the Gap: From Regular Expressions to Machines

Earlier I said that when regular expression libraries are implemented in programming languages, the way that they work is by converting regular expressions to finite state machines and then using the FSM that they generated to process input strings.

Doing the translation from the regular expression to an FSM is interesting. There are several different ways of converting a regular expression to an FSM. It's not a one-to-one translation. Depending on which method you choose, you can wind up with different FSMs for the same regular expression. (In fact, theoretically, there are an infinite number of different finite state machines for each regular expression!) Fortunately, that doesn't matter, because all of the different FSMs that you can generate for a particular regular expression will process exactly the same language and will do it in exactly the same amount of time. We're going to use the method that I think is the easiest to understand, called *Brzozowksi derivatives*, or simply the *derivative* of a regular expression.

The idea of the derivative is that you can look at a regular expression and ask, "If I give this regular expression one input character, what would it accept after that character?"

In formal terms, let $S(r)$ be the set of strings accepted by the regular expression r. Then if you have a regular expression r and a character c, the derivative of r with respect to c (written $D_c(r)$) is a regular expression r' such that $t \in S(r')$ if and only if $ct \in S(r)$.

For example, if you had a regular expression ab^*, then the derivative with respect to a would be ab^*.

If we know how to compute the derivative of a regular expression, then to convert from a regular expression, we can do the following:

1. Create an initial state, labeled with the complete regular expression r.
2. While there are states r_i in the machine that haven't been processed yet,
   a. For each character, c, in the alphabet, compute the derivative r'_i;
   b. If there is a state r'_i already in the machine, then add a transition from r_i to r'_i labeled with symbol c; and
   c. If there is no state r'_i, then add it and add a transition from r_i to r'_i labeled with the symbol c.

3. For each state in the machine labeled with a regular expression *r*, mark it as a final state if and only if *r* can match the empty string.

Computing the derivative is complicated, not because it's difficult but because there are a lot of cases to cover. I think that the easiest way to understand the process is to look at an implementation of it. We'll walk through it, looking at code that computes it in Haskell.

First, we need to declare how we're going to represent regular expressions. The Haskell code is very straightforward. A regular expression is defined exactly the way that we described it earlier: it's a specific character, a choice between multiple regular expressions, a sequence of regular expressions, or a repetition. For the Haskell, we'll add two alternatives that will make implementing the derivative easier: VoidRE, which is a regular expression that never matches anything, and Empty, which is a regular expression that only matches an empty string.

computing/deriv.hs

```
data Regexp = CharRE Char
            | ChoiceRE Regexp Regexp
            | SeqRE Rexexp Regexp
            | StarRE Regexp
            | VoidRE
            | EmptyRE
            deriving (Eq, Show)
```

To compute the derivative of a regular expression, we need to be able to test whether or not a given regular expression can accept the empty string. By convention, we call that function *delta*.

computing/deriv.hs

```
  delta :: Regexp -> Bool
❶ delta (CharRE c) = False
❷ delta (ChoiceRE re_one re_two) =
    (delta re_one) || (delta re_two)
❸ delta (SeqRE re_one re_two) =
    (delta re_one) && (delta re_two)
❹ delta (StarRE r) = True
❺ delta VoidRE = False
❻ delta EmptyRE = True
```

❶ A regular expression that only matches a specific character can't match the empty string.

2. A choice between two regular expressions can match an empty string if either (or both) of its alternatives can match an empty string.

3. One regular expression followed by a second regular expression can match an empty string only if *both* of the regular expressions can match the empty string.

4. A starred regular expression matches zero or one repetitions of a pattern. Zero repetitions is the empty string, so any starred regular expression can match the empty string.

5. The void-regular expression never matches anything, so it can't match the empty string.

6. By definition, the empty regular expression matches the empty string.

With delta out of the way, we can finally see how the derivative works!

computing/deriv.hs

```
derivative :: Regexp -> Char -> Regexp
derivative (CharRE c) d =                                            -- (1)
  if c == d
    then EmptyRE
    else VoidRE
derivative (SeqRE re_one re_two)) c =                                -- (2)
  let re_one' = (derivative re_one c)
  in case re_one' of
    VoidRE -> VoidRE
    EmptyRE -> re_two
    _ -> if (delta re_one)
           then (ChoiceRE (SeqRE re_one' re_two) (derivative re_two c))
           else (SeqRE re_one' re_two)
derivative (ChoiceRE re_one re_two) c =                              -- (3)
  let re_one' = (derivative re_one)
      re_two' = (derivative re_two)
  in case (re_one', re_two') of
    (VoidRE, VoidRE) -> VoidRE
    (VoidRE, nonvoid) -> nonvoid
    (nonvoid, VoidRE) -> nonvoid
    _ -> (ChoiceRE re_one' re_two')
derivative (StarRE r) c =                                            -- (4)
  let r' = derivative r c
  in case r' of
     EmptyRE -> (StarRE r)
     VoidRE -> VoidRE
     _ -> (SeqRE r' (StarRE r))
```

```
❺ derivative VoidRE c = VoidRE
  derivative EmptyRE c = VoidRE
```

❶ The derivative of a single-character pattern CharRE c with respect to a character k is *empty* if $k = c$, because the regular expression matched the character; otherwise it's a value representing failure, which we'll call *void*, because the match failed.

❷ The derivative of a sequence is the hardest thing to understand because there are a lot of subtle cases. We start by taking the derivative of the first regular expression. If that is void, that means that it can't possibly match, then the sequence can't match, so the derivative of the sequence is void as well. If the derivative of the first regular expression is empty, then it definitely matches and the derivative is just the second regular expression. If the derivative of the first is neither empty nor void, then there are two subtle choices: the obvious one is that the derivative of the sequence would be the derivative of the first regular expression followed by the second expression. But if the first regular expression could match an empty string, then we also need to consider the case where the first regular expression matches the empty string.

❸ This case is easy: the derivative of a choice is a choice between the derivatives of the alternatives.

❹ A starred regular expression is basically a choice between the empty string or a sequence consisting of a single instance of the starred expression followed by the star. That's hard to parse in English, but if we write it in a regular expression, $R^* = empty \mid (R(R^*))$.

❺ Both void and empty can't match a character, so their derivative with respect to any character must be *void*.

Let's try looking at an example of this process. We'll start with a regular expression that describes the same language as the machine in Figure 15, *The AB finite state machine*, on page 186: aa^*b^*:

1. The initial state of the machine will be labeled with the entire regular expression.

2. From the initial state, we need to take two derivatives, one with respect to *a* and one with respect to *b*:

    a. The derivative with respect to "a" is a^*b^*, so we add a state with that regular expression and connect the initial state to it with an arc labeled *a*.

    b. The derivative with respect to *b* is void because the regular expression doesn't allow a string to start with *b*. So we add a *void* state to the machine and an arc from the initial state to void labeled with *b*.

3. Now, we need to look at the state a^*b^*.

    a. The derivative of this regular expression with respect to *a* is the regular expression itself. We don't need to add a new state since we've already got this one. We just add an arc from the state to itself, labeled *a*.

    b. The derivative of this regular expression with respect to *b* is b^*, so we add a new state to the machine and an arc from the current state to the new state labeled with *b*.

4. We move on to the state b^*. For that, we'll get an arc labeled *a* to *void* and an arc labeled *b* from the state to the state itself.

5. Finally, we need to figure out which states are final. To do that, we need to compute *delta* for each state:

    a. *delta*(aa^*b^*) is false, so state aa^*b^* isn't final.

    b. *delta*(a^*b^*) is true, so it's a final state.

    c. *delta*(b^*) is true, so it's a final state.

    d. *delta(void)* is false, so it's not a final state.

The result is a machine that is identical to the machine we drew in the original figure, except that the state names are different.

Once you can compute the derivatives of regular expressions, it's easy to generate the FSM, and the FSM that you generate is an efficient way of processing strings. In fact, you can even use it to implement a lightweight regular expression matcher that doesn't need to generate the full FSM in advance! For each input symbol, you just take the derivative of the expression. If it's not the void expression, then go on to the next character, using the derivative to process the rest of the string. When you get to the end of your input, if δ of the final derivative regular expression is empty, then you accept the string.

If you do this intelligently, by doing something like memoizing the derivative function so that you're not constantly recomputing derivatives, this ends up being a reasonably efficient way to process regular expressions. (Memoization is a technique where you save the results of every invocation of a function, so that if you call it repeatedly with the same input, it doesn't redo the computation, but just returns the same result as the last time it was called with that input.)

That's the finite state machine, the simplest kind of computing machine. It's a really useful kind of machine in practice, and it's also genuinely interesting as an example of how machines work.

When we think about the computers that we use every day, we informally say that they're more powerful than a finite state machine. But in reality, that's not true. To be more powerful than an FSM, a machine needs to have an infinite (or at least unbounded) amount of storage, and the computers we use are obviously finite. As trivial as it seems, this is the reality of computers: since every real computer only has a finite amount of storage, computers are really all finite state machines. As finite state machines, they're *huge*: without considering disk storage, the computer that I'm using to write this book has about $2^{32,000,000,000}$ possible states! Because of the quantity of storage that they have, they're insanely complicated finite state machines. We don't normally think of them as incomprehensibly large FSMs, because we know that we can't understand an FSM with more possible states than there are particles in the known universe! Instead, we think of them as limited examples of a more powerful type of computing machine (like the Turing machine that we'll look at in the next chapter) that combine a small, comprehensible state machine with an unlimited amount of storage.

22

The Turing Machine

One of the greatest names in the history of math is Alan Turing (1912–1954). Turing was an amazing man who worked in a ton of different areas, most specifically in mathematical logic. At the heart of his most well-known work in computation is a theoretical machine that he designed as a model of mechanical computation, which is named the *Turing machine* in his honor.

The Turing machine isn't a model of real computers. The computer that I'm using to write this book has absolutely nothing to do with the Turing machine in any practical sense. As a real device, the Turing machine is absolutely terrible. But that's because it was never intended to be a real machine!

The Turing machine is a mathematical model not of computers but of *computation*. That's a really important distinction. The Turing machine is an easy-to-understand model of a computing device. It's definitely not the simplest model. There are simpler computing devices (for example, there's a cellular automaton called rule 111 that is dramatically simpler), but their simplicity makes them harder to understand. The Turing machine strikes a balance between simplicity and comprehensibility that is, at least in my opinion, completely unequalled.

The reason that the Turing machine is so important comes down to the fact that, theoretically, what machine you use to talk about computation doesn't matter. There's a limit to what a mechanical device can do. There are lots of machines out there, but ultimately no machine can go past the limit. Any machine that can reach that limit is, for the purposes of understanding computation, pretty much the same as any other. When we talk about studying computation, what we're talking about is the set of things that can be done by a machine—not by a particular machine but by *any* conceivable machine. The choice comes down to which machine makes things easiest to understand. And that's where the Turing machine stands out: it's remarkably easy to understand what it does, it's easy to tweak for experiments, and it's easy to use in proofs.

Adding a Tape Makes All the Difference

So let's take a step back and ask, what is a Turing machine?

A Turing machine is just an extension of a finite state machine. Just like an FSM, a Turing machine has a finite set of states and a state relation that defines how it moves between them based on its input. The difference is that its inputs come on a strip of tape, and the machine can both read *and write* symbols on that tape, something like this:

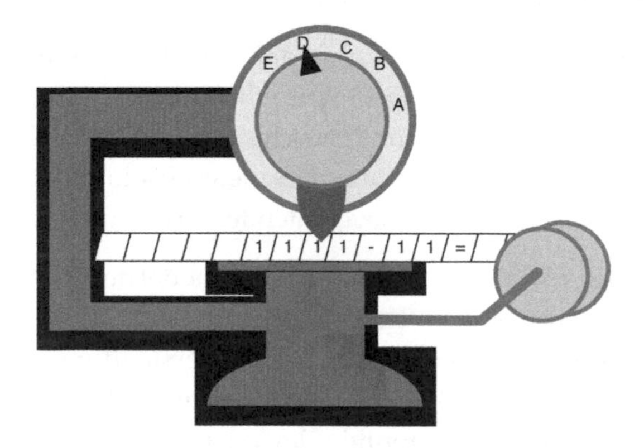

Figure 16— A Turing machine: *A Turing machine is a finite state machine that can read and write a tape.*

The basic idea of the Turing machine is simple. Take a finite state machine. Instead of feeding an input string directly into the machine, the way we did with the FSM, write it down onto a strip of tape. The tape is divided into cells, and each cell has one character on it. In order to use the tape, the machine has a *head* pointing at a cell, and it can either read or write the symbol on the tape. Like a finite state machine, the Turing machine will look at an input symbol and decide what to do. In the FSM, the only thing it could do was change states and go on to the next input character. With a Turing machine, it has the tape, so it can do more things. Each step, it looks at the tape cell under the head, and based on its state and the contents of the cell, it can change the state, change the current symbol on the tape, and move the tape head either left or right.

That's all that it takes to build a Turing machine. People who like to make computing sound impressive often have very complicated explanations of it, but really that's all there is to it: it's a finite state machine with a tape. The point of it was to be simple, and simple it certainly is. But the important fact is, if you have a task that can be done mechanically, it can be done by a Turing machine. Now we'll see how those pieces turn into a computer.

To really understand how that trivial machine can do computations, it helps to look at the formal definition of the machine and the way that it works. In formal mode, a Turing machine consists of the following pieces:

States A Turing machine has a set of *states*. At any time, the machine is in one of those states. Its behavior when it finds a particular symbol on the tape depends on the value of its current state. We'll use S for the set of states.

Another way to think of the state is as a small, fixed-size set of data that the machine can use to make decisions. But for the Turing machines we're going to look at, we'll always use a specific set of states. (You'll see what I mean in a minute.)

There's one specific state, called the *initial state*. When the machine starts running, before it's looked at the tape or done anything, it's in its initial state.

To tell when the machine is done running, there's a second special state called the *halting state*. When the machine enters the halting state, it stops, and whatever is on the tape is the result of the computation.

Alphabet Each machine has a collection of symbols that it can read from and write to its tape. This set is called the machine's *alphabet*.

Transition Function This is the real heart of the machine, which describes how the machine behaves. For formality, it's defined as a function from the machine state and the alphabet character on the current tape cell to the action that the machine should take. The action specifies a new value for the machine's state, a character to write onto the current tape cell, and a direction to move the tape head, either left or right.

For example, let's look at a classic example of a really simple Turing machine: one that does subtraction using *unary* numbers. A unary number *N* is written as a series of *N* "1"s. In unary notation, for example, the number 4 is written as 1111.

We'll give the machine a tape that contains the two numbers *M* and *N* to be subtracted, written as a string "*N* - *M* =." After running until it halts, the tape will contain the value of *M* subtracted from *N*. For example, if the input tape contains the characters "1111-11=" (or 4 - 2 in decimal), the output will be "11" (or 2 in decimal).

The alphabet is the characters "1", " " (blank space), "-" (minus sign), and "=" (equals sign).

The machine has four states: *scanright*, *eraseone*, *subone*, and *skip*. It starts in the state *scanright*. Its transition function is given by the following table:

FromState	Symbol	ToState	WriteChar	Direction
scanright	space	scanright	space	right
scanright	1	scanright	1	right
scanright	minus	scanright	minus	right
scanright	equal	eraseone	space	left
eraseone	1	subone	equal	left
eraseone	minus	HALT	space	n/a
subone	1	subone	1	left
subone	minus	skip	minus	left
skip	space	skip	space	left
skip	1	scanright	space	right

What this machine does is move to the right until it sees the equals sign; it erases the equals sign and moves to the left, erases one digit off the second number and replaces it with the equals sign (so the second number is reduced by one and the equals sign is moved over one position). Then it scans back to the minus sign (which separates the two numbers), erases one digit of the first number, and switches back to scanning to the right for the equals sign.

So one at a time it erases one digit from each of the two numbers. When it reaches the equals sign, it starts going back to erase a digit from the second number; if it hits the "-" before it finds a digit, it knows that it's done, so it stops. That will be clearer if we trace through how the machine would process a specific input string. In the trace, I'll write the machine state, followed by a colon, followed by the tape contents surrounded by [] (square brackets), with the current tape cell surrounded by {} (curly braces).

State	Tape
scanright	[{1}1 1 1 1 1 1 1 - 1 1 1 =]
scanright	[1{1}1 1 1 1 1 1 - 1 1 1 =]
scanright	[1 1{1}1 1 1 1 1 - 1 1 1 =]
scanright	[1 1 1{1}1 1 1 1 - 1 1 1 =]
scanright	[1 1 1 1{1}1 1 1 - 1 1 1 =]
scanright	[1 1 1 1 1{1}1 1 - 1 1 1 =]
scanright	[1 1 1 1 1 1{1}1 - 1 1 1 =]
scanright	[1 1 1 1 1 1 1{1}- 1 1 1 =]
scanright	[1 1 1 1 1 1 1 1{-}1 1 1 =]
scanright	[1 1 1 1 1 1 1 1 -{1}1 1 =]
scanright	[1 1 1 1 1 1 1 1 - 1{1}1=]
scanright	[1 1 1 1 1 1 1 1 - 1 1{1}=]
scanright	[1 1 1 1 1 1 1 1 - 1 1 1{=}]
eraseone	[1 1 1 1 1 1 1 1 - 1 1{1}]
subone	[1 1 1 1 1 1 1 1 - 1{1}=]
subone	[1 1 1 1 1 1 1 1 -{1}1=]
subone	[1 1 1 1 1 1 1 1{-}1 1=]
skip	[1 1 1 1 1 1 1{1}-1 1=]

State	Tape
scanright	[1 1 1 1 1 1 1 {-}1 1=]
scanright	[1 1 1 1 1 1 1 -{1}1=]
scanright	[1 1 1 1 1 1 1 - 1{1}=]
scanright	[1 1 1 1 1 1 1 - 1 1{=}]
eraseone	[1 1 1 1 1 1 1 - 1{1}]
subone	[1 1 1 1 1 1 1 -{1} =]
subone	[1 1 1 1 1 1 1{-}1 =]
skip	[1 1 1 1 1 1 1{ }- 1 =]
skip	[1 1 1 1 1 1{1} - 1 =]
scanright	[1 1 1 1 1 1 { }- 1 =]
scanright	[1 1 1 1 1 1 {-}1 =]
scanright	[1 1 1 1 1 1 -{1} =]
scanright	[1 1 1 1 1 1 -1{=}]
eraseone	[1 1 1 1 1 1 -{1}]
subone	[1 1 1 1 1 1 {-}=]
skip	[1 1 1 1 1 1 { }- =]
skip	[1 1 1 1 1 1{ } - =]
skip	[1 1 1 1 1{1} - =]
scanright	[1 1 1 1 1 { } - =]
scanright	[1 1 1 1 1 { }- =]
scanright	[1 1 1 1 1 {-}=]
scanright	[1 1 1 1 1 -{=}]
eraseone	[1 1 1 1 1 {-}]
Halt	[1 1 1 1 1 { }-]

The result is 11111 (5 in decimal). See—it works!

One really important thing to understand here is that we *do not have a program*. What we just did was define a Turing machine that does subtraction. This machine *does not* take any instructions: the states and the transition function are built into the machine. The only thing this specific Turing machine can do is subtract one number from another. To add or multiply two numbers, we would have to build another machine.

Going Meta: The Machine That Imitates Machines

Turing machines like the ones we've seen are nice, but they're very limited. If all Turing had done was invent a machine like this, it would have been cool but not really remarkable. The real genius of Turing was his realization that this kind of machine was enough *to be able to imitate itself*. Turing was able to design a Turing machine whose input tape contained a description of another Turing machine—what we would now call a program. This single machine, known today as a *universal Turing machine,* could simulate *any* other Turing machine and therefore could be programmed to perform any computation!

That's the basic idea of what we call a computer, and it tells us what a computer program really is. The universal Turing machine (or UTM) isn't just a computing machine: it is a device that can be turned into *any* computing machine just by feeding it a description of the machine you want. And a computer program, whether it's written in binary machine language, lambda calculus, or the latest functional programming language, is nothing but a description of a machine that does a specific task.

The universal Turing machine isn't just a machine: it's a machine that is able to masquerade as other machines. You don't need to build special machines for special tasks: with the Turing machine, you only need one machine, and it has the ability to become any other machine that you want.

To understand computing, we play with the Turing machine, using it as a platform for experiments. Figuring out how to do things in terms of the simple mechanics of a Turing machine can be a fascinating exercise: nothing else drives home quite how simple of a machine this really is.

A great example of Turing machine experimentation involves the effort to figure out what the smallest possible UTM is. For years, people figured out how to make the machine smaller and smaller: seven states and four symbols, five states and four symbols, four states and three symbols. Finally, in 2007, a two-state machine with a three-symbol alphabet (first proposed in *A New Kind of Science [Wol02]*) was shown to be "Turing complete." It's been known for a while that it's not possible to build a UTM with fewer than three symbols in its alphabet, so the two/three machine is now known to be the simplest possible.

Another set of experiments is possible when you start trying to change the machine and see what effect it has. When you see how simple the Turing machine is, it seems hard to believe that it really *is* a universal machine. What we can do to probe that is to try adding things to the machine and then see if they let us do anything that we couldn't do with a standard Turing machine.

For example, if you try to do complex computations with a Turing machine, you end up spending a lot of time scanning back and forth, trying to find the positions where you need to do something. We saw this in our simple subtraction example: even in something as trivial as unary subtraction, it took some effort to get the forward and backward scanning right. What if we added a second tape to be used for auxiliary storage? Would we be able to do anything that we couldn't do with a single-tape Turing machine?

Let's get specific about what we want to do. We're going to create a new two-tape Turing machine. The input to the machine is on the first tape, and the second tape will be blank when the machine starts. As it scans the tape, it can write markings on the second tape to record information about the computation in progress. The tapes move together so that the annotations on the second tape are always lined up with the things that they annotate. The transition function is extended to two tapes by having each state transition rule depend on the pair of values found on the two tapes, and it can specify symbols to be written onto both tapes.

Does the auxiliary storage tape add any power? No, you can design a single-tape Turing machine that behaves exactly the same as the two-tape. You just need to be able to write two symbols onto each tape cell, and you can do that by creating a new alphabet, where each alphabet symbol actually consists of a pair. Take the set of symbols that you can have on tape 1 and call it A_1, and take the set of symbols you can have on tape 2 and call that A_2. We can create a single-tape Turing machine whose alphabet is the cross-product of A_1 and A_2. Now each symbol on the tape is equivalent to a symbol on tape 1 and a symbol on tape 2. So we've got a single-tape machine that is equivalent to the two-tape machine. With that change, the single-tape Turing machine can do exactly the same thing as the two-tape. And if the fancy augmented-alphabet Turing machine can do it, since it's a regular one-tape Turing machine, that means that a universal Turing machine can do it as well!

We could do a lot more. For example, we could lift the restriction on the heads moving together. A two-tape machine where the tapes move independently is much more complicated. But you can still show how a single-tape machine can do the same things. The two-tape machine can be *much* faster at a lot of computations. To emulate the two tapes, a one-tape machine is going to have to do a lot of scanning back and forth between the positions of the two heads. So a single-tape simulation of a two-tape machine is going to be a whole lot slower. But while it might take it a lot longer, anything that you can do with a two-tape machine, you can do with a one-tape.

How about a two-dimensional tape? There are some fun programming languages that are based on the idea of a two-dimensional Turing machine.[1] It certainly seems like there's a lot of expressibility in the two dimensions. But as it turns out, there's nothing doable in two dimensions that isn't doable in one!

We can simulate a two-dimensional machine with a two-tape machine. Since we know that we can simulate a two-tape with a one-tape, if we can describe how to do a two-dimensional machine with a two-tape machine, we'll know that we could do it with just one tape.

For a two-tape machine, we map the 2D tape onto the 1D-tape, as seen in Figure 17, *Mapping a 2D tape*, on page 206, so that cell 0 on the 1D tape corresponds to cell (0,0) on the two-tape, cell (0,1) on the 2D corresponds to cell 1 on the 1D, cell (1,1) on the 2D is cell 2 on the 1D, and so on. Then we use

1. The esoteric language Befunge (http://catseye.tc/node/Funge-98.html) is a fun example of programming a two-dimensional universal Turing machine.

the second tape for the bookkeeping necessary to do the equivalent of 2-D tape moves. And we've got a 2D Turing machine simulated with a two-tape 1D; and we know that we can simulate a two-tape 1D with a one-tape 1D.

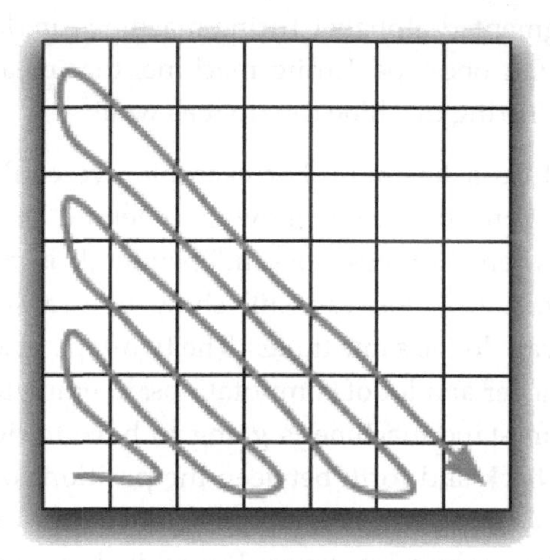

Figure 17—Mapping a 2D tape

To me, this is the most beautiful thing about the Turing machine. It's not just a fundamental theoretical construction of a computing device; it's a simple construction of a computing device that's really easy to experiment with. Consider lambda calculus for a moment. It's more useful than a Turing machine for lots of purposes. In the real world, we write programs in lambda calculus when no one would build a real application using a Turing machine program. But imagine how you'd try things like the alternate constructions of the Turing machine. It's a whole lot harder to build experiments like those in lambda calculus: likewise for other kinds of machines, like Minsky machines, Markov machines, and so forth.

If you're interested in playing with a Turing machine, I implemented a simple Turing machine language in Haskell. You can get the source code and compilation instructions for it on the website for this book.[2] You feed it a Turing

2. http://pragprog.com/book/mcmath/good-math

machine description and an input string, and it will give you a trace of the machine's execution like the one discussed. Here's the specification of the subtraction machine written in my little Turing language:

```
states { "scanright" "eraseone" "subtractOneFromResult"
    "skipblanks" } initial "scanright"
alphabet { '1' ' ' '=' '-' } blank ' '
trans from "scanright" to "scanright" on (' ')
    write ' ' move right
trans from "scanright" to "scanright" on ('1')
     write '1' move right
trans from "scanright" to "scanright" on ('-')
     write '-' move right
trans from "scanright" to "eraseone" on ('=')
    write ' ' move left
trans from "eraseone" to "subtractOneFromResult" on ('1')
    write '=' move left
trans from "eraseone" to "Halt" on ('-')
    write ' ' move left
trans from "subtractOneFromResult" to
    "subtractOneFromResult" on ('1')
    write '1' move left
trans from "subtractOneFromResult" to "skipblanks" on ('-')
    write '-' move left
trans from "skipblanks" to "skipblanks" on (' ')
    write ' ' move left
trans from "skipblanks" to "scanright" on ('1')
    write ' ' move right
```

The syntax is pretty simple:

- The first line declares the possible states of the machine and what state it starts in. This machine has four possible states: "scanright," "eraseone," "subtractOneFromResult," and "skipblanks." When the machine starts, it will be in the "skipright" state.
- The second line declares the set of symbols that can appear on the tape, including what symbol will initially appear on any tape cell whose value wasn't specified by the input. For this machine, the symbols are the digit 1, a blank space, the equals sign, and the subtraction symbol; the blank symbol is on any tape cell whose initial value wasn't specified.
- After that is a series of transition declarations. Each declaration specifies what the machine will do for a given pair of an initial state and a tape symbol. So, for

example, if the machine is in state "scanright" and the current tape cell contains the equals sign, then the machine will change to state "eraseone," write a blank onto the tape cell (erasing the "=" sign), and move the tape cell one position to the left.

That's the machine that changed the world. It's not a real computer, and it's got very little to do with the computer that's sitting on your desk, but it's the one that laid the groundwork not just for our computers but for the entire concept of computation. As simple as it is, there's nothing that you can do with any computer that you can't do with a Turing machine.

The fundamental lesson of the Turing machine is that computation is *simple*. It doesn't take much to make it work. In the next chapter, we'll take a look at computation from a different, mind-warping direction; and in the process, we'll explore just what a computing system needs in order to be able to do any possible computation, just like a Turing machine.

23

Pathology and the Heart of Computing

In computer science–speak, we say that a computing system is "Turing complete" if it can do the same computations as a Turing machine. That's important because the Turing machine is an example of the maximum capability of any mechanical computer. The capability of a Turing machine can be matched, but never exceeded. If a computing device can do the same computations as a Turing machine, then it can do everything that any computer can do. Understanding that, we'd like to understand what a machine needs to be Turing complete. How hard is it to make machines that can perform any possible computation?

The answer is, surprisingly easy.

Computers can do amazingly complicated things, and because of that, we expect that they must be complicated. Our experiences with these devices seems to support our expectation of complexity: the computer on which I'm typing has 292 *million* switches in its CPU alone, and it has billions more in RAM, flash storage, its graphics processing unit, and elsewhere. That's the kind of number that's really hard to wrap your head around. A modern solid-state silicon computer is an amazingly complex machine.

But despite their complexity and despite our expectations, computing devices are, in theory at least, extremely simple machines. What's complicated is figuring out how to make them small and fast, how to make them easy to program, or how to build them so that they can interact with lots of devices. But in fact, the basics of computation are so simple that you have to work hard to build a computing system that *isn't* Turing complete.

So what does it take to be Turing complete? There are four essential elements that form the heart of computing: any machine that has all four will be Turing complete and therefore capable of performing any possible computation.

- *Storage*: Every complete computing device needs to have access to an unbounded amount of storage. Obviously, no real machine can have infinite storage, and no program can ever use an infinite amount of storage. But in theory, to be Turing complete, you can't have any fixed limits on the amount of storage you can access. The storage can be anything you want. It doesn't have to be a numerically addressable store, like what we have in real computers. It could be a tape, or a queue, or a collection of name-to-value bindings, or an expandable collection of grammar rules. It doesn't matter what kind of storage it is, as long as it's unlimited.

- *Arithmetic*: You need to be able to do arithmetic in some way. In particular, you need to be able to do Peano arithmetic. The specifics don't really matter. You can do your arithmetic in unary, binary, ternary, decimal, EBCDIC, roman numerals, or anything else you can imagine. But your machine must *at least* be able to perform the basic operations defined by Peano arithmetic.

- *Conditional Execution*: To do general computation, you need to have some way of making choices. You can do that by selectively ignoring certain code (like the PLEASE IGNORE command in INTERCAL), by having conditional branches, by having state selection based on the contents of a tape cell, or by a variety of other mechanisms. The key is that you need to have the ability to select different behaviors based on values that were either computed or were part of the input to your program.

- *Repetition*: Every computing system needs to be able to repeat things. Loops, recursion, or some other repetition for iterating or repeating pieces of a computation are essential, and they need to be able to work together with the mechanism for conditional execution to allow conditional repetition.

In this chapter, we're going to use those requirements to study how a new computing machine provides the essential requirements of a Turing-complete computing system. But just looking at another machine like the Turing machine would be boring. So we're going to go the silly route and look at the heart of computing by looking at what I call a *pathological programming language*.

Introducing BF: The Great, the Glorious, and the Completely SIlly

In real life, I'm not a mathematician; I'm a computer scientist. I'm still a math geek, mind you, but what I really do is very much in the realm of applied math, working on building systems to help people build programs.

One of my pathological obsessions is programming languages. Since I first got exposed to TRS-80 Model 1 BASIC back in middle school, I've been absolutely nuts about programming languages. Last time I counted, I'd learned about 150 different languages, and I've picked up more since then. I've written programs in most of them. Like I said, I'm nuts.

This is a roundabout way of explaining where the subject of this chapter came from. There's a remarkably simple computing device, with a complete formal definition and an implementation in hardware,[1] that has been rendered into an amazingly bizarre pathological programming language.

Designing bizarre programming languages is a popular hobby among a certain wonderfully crazy breed of geeks. There are, at the very least, hundreds of these languages.[2] Even in a gallery full of insane programming languages,

1. http://en.wikipedia.org/wiki/P_prime_prime and http://www.robos.org/?bfcomp, respectively
2. If you're interested, there's a wiki full of them at http://esolangs.org/.

there's one that deserves a special place of honor: Brainf***,[3] designed by a gentleman named Urban Möller.

The BF machine is a work of beauty. It's a *register machine,* which means that its computations are performed in storage registers. If you're familiar with a modern electronic CPU, you're familiar with a real-life register machine. In real computer hardware, the physical hardware registers have fixed names. In the BF machine they don't: they're referenced via relative position. On a BF machine, an infinite number of registers are (conceptually) available on an infinite tape, and there's a tape head that points to the current register. Registers are referenced by their position relative to the current position of the tape head. (BF makes the concept of relative addressing, a technique used by most modern compilers, into a fundamental primitive.) BF programs work by moving the register tape head back and forth to access different registers. When the register tape head is located at a particular cell, it can increment the cell's value, decrement the cell's value, and branch conditionally, depending on whether the cell's value is zero. That's it, and that's enough to be Turing complete.

Now that we know a bit about the machine, let's look at its instruction set. BF has a total of eight commands, including input and output. Each command is written as a single character, making BF among the most concise language syntaxes ever.

Tape Forward: > Move the tape head one cell forward

Tape Back: < Move the tape head one cell backward

Increment: + Increment the value on the current tape cell

Decrement: - Decrement the value on the current tape cell

Output: . Output the value on the current tape cell as a character

Input: , Input a character and write its numeric value onto the current tape cell

3. http://www.muppetlabs.com/~breadbox/bf/

Compare and Branch Forward: [Compare and jump forward —compare the current tape cell to zero: if it's zero, jump forward to the first instruction after the matching "]"; otherwise, go on to the next instruction.

Compare and Branch Back:] Compare and jump backward—if the current tape cell is not zero, then jump backward to the matching "[".

BF ignores any character that is not one of those eight instruction characters. When the BF interpreter encounters a non-command character, it will just skip to the next command character. This means that you don't need any special syntax to write comments in BF, because you can simply intersperse your comments with the program instructions. (But you need to do it carefully; if you use punctuation, you'll probably create instructions that will break your program.)

To give you some sense of what it looks like, here's a hello-world program in BF:

```
++++++++
[>+++++++++<-]
>.<+++++
[>++++++<-]
>-.+++++++..+++.
<++++++++
[>>++++<<-]
>>.<<++++
[>------<-]
>.<++++
[>++++++<-]
>.+++.------.
--------.>+.
```

Let's pull that program apart in an attempt to understand it.

- ++++++++: Store the number 8 in the current tape cell. We're going to use that as a loop index, so the loop is going to repeat eight times.
- [>+++++++++<-]: Run a loop: using the tape cell after the loop index, add 9 to it. Then go back to the loop index, decrement it, and, if it's not zero, branch back to the beginning of the loop. When the loop completes,

we'll wind up with the number 72 in the second cell. That's the ASCII code for the letter *H*.

- >.: Go to the cell after the loop index and output what's there. That outputs the value 72 as a character (*H*).
- <+++++: Return to the loop index. This time store the value 5 there.
- [>++++++<-]: Run another loop, as we did to generate the *H*: this time, however, we're going to add 6 to the value in the second cell 5 times. (Remember that we didn't get rid of the value left in that cell by our previous operation, which is still the number 72.) When this new loop completes, the second cell contains the number 102.
- >-.: Advance past the index, subtract 1, and output the value in the register, which is the number 101, or *e*.

After that, it continues in pretty much the same vein, using a couple of tape cells and running loops to generate the values of the characters. It's quite beautiful in its way. But at the same time, that's an astonishingly complicated way of just printing out "Hello world"! Remarkable, isn't it?

Turing Complete, or Completely Pointless?

So how is the BF machine Turing complete? Let's look at its features in terms of the four criteria for being Turing complete.

- *Storage:* BF has an unbounded tape. Each cell on that tape can hold an arbitrary integer value. So the storage is obviously unbounded. It's tricky to work with, because you can't reference a specific cell by name or by address: the program has to be written to keep track of where the tape head currently is and how to move it to get to the location of the value you want to look at. But when you think about it, that's not really a restriction. In a program on a real computer, you need to keep track of where everything is—and in fact, most programs are written to use relative addresses anyway—so the BF mechanism, while incredibly simple, isn't particularly restrictive, certainly not enough to make this storage unusable for computing. So the storage requirement is satisfied.

- *Arithmetic:* BF stores integer values in its tape cells and provides increment and decrement operations. Since Peano arithmetic defines its operations in terms of increment and decrement, this is obviously sufficient. So we've got arithmetic.

- *Conditional Execution and Repetition:* BF provides both conditional execution and repetition through a single mechanism. Both the *[* and *]* operations provide conditional branching. The *[* can be used to conditionally skip over a block of code and as a target for a backward branch by a *[*. The *]* command can be used to conditionally repeat a block of code, and as a target for a forward branch by a *[* instruction. Between them you can create blocks of code that can be conditionally executed and repeated. That's all that we need.

From the Sublime to the Ridiculous

If that didn't seem impressive enough, here is a really gorgeous implementation of a Fibonacci sequence generator, complete with documentation. The BF compiler used to write this ignores any character other than the eight commands, so the comments don't need to be marked in any way; they just need to be really careful not to use punctuation.[4]

```
+++++++++++ number of digits to output
> #1
+ initial number
>>>> #5
++++++++++++++++++++++
++++++++++++++++++++++ (comma)
> #6
++++++++++++++++
++++++++++++++++ (space)
<<<<<< #0
[
> #1
copy #1 to #7
[>>>>>>+>+<<<<<<<-]
>>>>>>>[<<<<<<<+>>>>>>>-]
<
divide #7 by 10 (begins in #7)
[
>
```

4. The original source of this code is the BF documentation at http://esoteric.sange.fi/brainfuck/bf-source/prog/fibonacci.txt.

```
++++++++++  set the divisor #8
[
subtract from the dividend and divisor
-<-
if dividend reaches zero break out
copy dividend to #9
[>>+>+<<<-]>>>[<<<+>>>-]
set #10
+
if #9 clear #10
<[>[-]<[-]]
if #10 move remaining divisor to #11
>[<<[>>>+<<<-]>>[-]]
jump back to #8 (divisor position)
<<
]
if #11 is empty (no remainder) increment the quotient #12
>>> #11
copy to #13
[>>+>+<<<-]>>>[<<<+>>>-]
set #14
+
if #13 clear #14
<[>[-]<[-]]
if #14 increment quotient
>[<<+>>[-]]
<<<<<<< #7
]
quotient is in #12 and remainder is in #11
>>>>> #12
if #12 output value plus offset to ascii 0
[+++++++++++++++++++++++
+++++++++++++++++++++++++.[-]]
subtract #11 from 10
++++++++++  #12 is now 10
< #11
[->-<]
> #12
output #12 even if it's zero
++++++++++++++++++++++++
++++++++++++++++++++++++.[-]
<<<<<<<<<<< #1
check for final number
copy #0 to #3
<[>>>+>+<<<<-]>>>>[<<<<+>>>>-]
<- #3
if #3 output (comma) and (space)
[>>.>.<<<[-]]
<< #1
[>>+>+<<<-]>>>[<<<+>>>-]
<<[<+>-]>[<+>-]<<<-
]
```

Now this might seem pointless. I would honestly never recommend BF for writing serious programs. But BF is an amazingly simple language, and it's the best example that I know of a minimal Turing-complete computing system: eight instructions, *including* input and output.

It's a great demonstration of what it means to be Turing complete: BF has exactly the essentials that it needs: unbounded storage on the register tape, arithmetic by the increment and decrement instructions, and control flow and loops using the open and close bracket instructions. That's it: that's the heart of computing.

24

Calculus: No, Not *That* Calculus—λ Calculus

In computer science, especially in the field of programming languages, when we're trying to understand or prove facts about computation, we use a tool called λ ("lambda") calculus.

λ calculus was designed by an American mathematician named Alonzo Church (1903–1995) as part of one of the first proofs of the Halting problem (which we'll talk more about in 27, *The Halting Problem*, on page 253). Turing is largely credited as the person who solved that, and his proof is the one that most people remember. But Church did it independently of Turing, and his proof was actually published first!

λ calculus is probably the most widely used theoretical tool in computer science. For example, among programming language designers, it's the preferred tool for describing how programming languages work. Functional programming languages like Haskell, Scala, and even Lisp are so strongly based in λ calculus that they're just alternative syntaxes for pure λ calculus. But the influence of λ calculus isn't limited to relatively esoteric functional languages. Python and Ruby both have strong λ-calculus influences, and even C++ template metaprogramming is profoundly influenced by λ calculus.

Outside of the world of programming languages, λ calculus is extensively used by logicians and mathematicians studying the nature of computation and the structure of discrete mathematics; and it's even used by linguists to describe the meaning of spoken languages.

What makes it so great? We'll see in detail in this section, but the short version is this:

- λ calculus is simple: it's only got three constructs for building expressions—abstraction, identifier reference, and application. To evaluate expressions written using those two constructs, you only need two evaluation rules, called α (renaming ["alpha"]) and β (function application ["beta"]).
- λ calculus is Turing complete: if a function can be computed by any possible computing device, then it can be written in λ calculus.
- λ calculus is easy to read and write: it has a syntax that looks like a programming language.
- λ calculus has strong semantics: it's based on a logical structure with a solid formal model, which means that it's very easy to reason about how it behaves.
- λ calculus is flexible: its simple structure makes it easy to create variants for exploring the properties of various alternative ways of structuring computations or semantics.

λ calculus is based on the concept of *functions*. The basic expression in λ calculus is the definition of a function in a special form, called a λ expression. In pure λ calculus, everything is a function; the only thing you can do is define and apply functions, so there are no values except for functions. As peculiar as that may sound, it's not a restriction at all: we can use λ calculus functions to create any data structure we want.

With the lead-in out of the way, let's dive in and look at λ calculus.

Writing λ Calculus: It's Almost Programming!

Before we really dive in to the specifics of λ calculus, let's consider exactly what makes it a calculus. To most of us, the word *calculus* means something very specific: the differential and integral calculus invented by Isaac Newton and

Gottfried Leibnitz. λ calculus has *nothing* to do with that kind of calculus!

In mathematical terms, a *calculus* is a system for symbolically manipulating expressions. Differential calculus is a calculus because it's a way of manipulating expressions that represent mathematical functions. λ calculus is a calculus because it describes how to manipulate expressions that describe *computations*.

The way that we define a calculus is very much like the way that we defined a logic back in 12, *Mr. Spock Is Not Logical*, on page 79. It has a *syntax*, which describes how you write statements and expressions in the language, and it has a set of *evaluation rules* that allow you to symbolically manipulate expressions in the language.

One of the many reasons that I love λ calculus is because it's so natural to a programmer. While Turing machines are beautifully simple, trying to figure out how to do complicated things with one can be quite a challenge. But λ calculus? It's just like programming. You can understand it as being a template for building programming languages (in part, because it has become the main template for building programming languages). The language of λ calculus is basically a super-simple expression-based programming language with just three types of expressions:

- *Function definition:* A function in λ calculus is an expression, written as *λ param . body*, which defines a function with one parameter.

- *Identifier reference:* An identifier reference is a name that matches the name of a parameter defined in a function expression enclosing the reference.

- *Function application:* Applying a function is written by putting the function value in front of its parameter, as in $x\ y$ to apply the function x to the value y.

If you're paying attention, you might have noticed a problem with function definitions. They only allow one parameter! How can we write all of our functions with one parameter? We can't even implement a simple addition function using only one parameter!

The solution is one of the fundamental lessons of λ calculus: many things that we believe to be essential primitives aren't. λ calculus shows us that we don't actually need to have multiparameter functions, numbers, conditionals, or any of the other things that we tend to think of as basic primitives. We don't need any of those concepts as primitives because simple one-parameter functions are powerful enough to allow us to *build* them.

The lack of multiple-parameter functions isn't a problem: we can create things that behave just like multiple-parameter functions but that are built using nothing but single-parameter functions. This works because of the fact that in λ calculus, functions are values, and just like we create new values whenever we want in a program, we can create new functions whenever we want. We can use that ability to create the effect of a multiple-parameter function. Instead of writing a two-parameter function, we can write a one-parameter function that returns a one-parameter function, which can then operate on the second parameter. In the end, it's effectively the same thing as a two-parameter function. Taking a two-parameter function and representing it by 2 one-parameter functions is called *currying,* after the logician Haskell Curry (1900–1982), who originally thought of the concept.

For example, suppose we wanted to write a function to add x and y. We'd like to write something like this: $\lambda\ x\ y\ .\ x + y$. We do that with one-parameter functions like this:

- We write a function that takes the first parameter.
- The first function returns a second one-parameter function, which takes the second parameter, and returns a final result.

Adding x plus y becomes writing a one-parameter function with parameter x, which returns another one-parameter function with parameter y and which actually returns the sum of x and y: $\lambda\ x\ .\ (\lambda\ y\ .\ x + y)$. In fact, if we gave that a name, like "add," then we'd invoke it as *add* 3 4. The curried function even looks like a two-parameter function when we use it!

Because of currying, taking multiple parameters really isn't fundamentally different from taking just one parameter, so long as you can create and return new functions. (See what I mean about λ calculus being great for experimentation? We've barely looked at any of it, and it's already coming in handy!)

In practice, we'll go ahead and write λ expressions with multiple parameters. It's just a simplified syntax for the curried functions, but it's very convenient, and it can make expressions easier to read.

There's one other really important thing about understanding λ calculus that I haven't mentioned yet. Look at our second currying example. It will work correctly only if, when it returns the function $\lambda\, y\, .\, x + y$, the variable x gets its value from the invocation of the λ form that *textually* surrounds it. If it's a standalone and x can get its value from anywhere but the invocation of the surrounding λ, then it won't produce the right result.

That property, that variables are always bound by their specific textual context, is called *syntactic closure* or *syntactic binding*. In programming languages, we call it *lexical binding*. It's how we tell which definition of a variable we'll use in a function: no matter where a function is *used*, the values of all of the variables that it uses take their meaning from where it was *defined*.

Like many programming languages, every variable in λ calculus must be declared. The only way to declare a variable is to *bind* it using a λ expression. For a λ calculus expression to be evaluated, it cannot reference any identifiers that are not bound. An identifier is bound if it is defined as a parameter in an enclosing λ expression; if an identifier is not bound in any enclosing context, then it is called a *free* variable. Let's look quickly at a few examples:

- $\lambda\, x\, .\, p\ x\ y$: In this expression, y and p are free because they're not the parameter of any enclosing λ expression; x is bound because it's a parameter of the function definition enclosing the expression $p\ x\ y$ where it's referenced.

- $\lambda\ x\ y.y\ x$: In this expression both x and y are bound because they are parameters of the function definition, and there are no free variables.
- $\lambda\ y\ .\ (\lambda\ x\ .\ p\ x\ y)$: This one is a tad more complicated because it contains an inner λ. So let's start there. In the inner λ, $\lambda\ x\ .\ p\ x\ y$, y and p are free and x is bound. In the full expression, both x and y are bound: x is bound by the inner λ, y is bound by the other λ, and p is still free.

We'll often use *free(x)* to mean the set of identifiers that are free in the expression x.

A λ calculus expression is valid (and thus evaluable) only when all of its variables are bound. When we look at smaller subexpressions of a complex expression taken out of context, they can look like they have free variables. That means that it's very important to have a way of making sure that the variables that are free in subexpressions are treated correctly. We'll see how we can do that by using a renaming operation called α in the next section.

Evaluation: Run It!

There are only two real rules for evaluating expressions in λ calculus, called *α conversion* and *β reduction*.

α is a renaming operation. In λ calculus, the names of variables don't have any meaning. If you rename a variable at its binding point in a λ and also rename it in all of the places where it's used, you haven't changed anything about its meaning. When you're evaluating a complex expression, you'll often end up with the same name being used in several different places. α conversion replaces one name with another to ensure that you don't have name collisions.

For instance, if we had an expression like $\lambda\ x\ .\ if\ (=x\ 0)\ then\ 1\ else\ x^2$, we could do an α conversion to replace x with y (written $\alpha[x/y]$) and get $\lambda\ y\ .\ if\ (=y\ 0)\ then\ 1\ else\ y^2$.

Doing an α conversion doesn't change the meaning of the expression in any way. But as we'll see later, it's important because without it, we'd often wind up with situations where a single variable symbol is bound by two different enclosing λs. (This will be particularly important when we get to recursion.)

β reduction is where things get interesting: this single rule is all that's needed to make the λ calculus capable of performing any computation that can be done by a machine.

A β reduction is how you apply a function in λ calculus. If you have a function application, you can apply it by replacing the function with the body of the λ and then taking the argument expression and replacing all uses of the parameter in the λ with the argument expression. That sounds confusing, but it's actually pretty easy when you see it in action.

Suppose we have this application expression: $(\lambda\, x\, .\, x + 1)\ 3$. By performing a β reduction, we can replace the application by taking the body $x + 1$ of the function and substituting (or αing) the value of the parameter (3) for the parameter variable symbol (x). (This is sometimes written as $\alpha[x/3]$.) So we replace all references to x with 3. So the result of doing a β reduction is $3 + 1$.

A slightly more complicated example is the expression $\lambda\, y\, .\, (\lambda\, x\, .\, x + y))\ q$. It's an interesting expression because it's a λ expression that, when applied, results in another λ expression; that is, it's a function that creates functions. When we do β reduction in this, we're replacing all references to the parameter y with the identifier q, so the result is $\lambda\, x\, .\, x + q$.

One more example, just for the sake of being annoying. Suppose we have $(\lambda\, x\, y.\ x\ y)\ (\lambda\, z\, .\, z * z)\ 3$. That's a function that takes two parameters and applies the first one to the second one. When we evaluate that, we replace the parameter x in the body of the first function with $\lambda\, z\, .\, z * z$, and we replace the parameter y with 3, getting $(\lambda\, z\, .\, z * z)\ 3$. And we can perform β on that, getting $3 * 3$.

Written formally, β says this:

$$\lambda x.Be = B[x/e] \text{ if free}(e) \subseteq \text{free}(B[x/e])$$

That condition on the end is why we need α: we can only do β reduction if doing it doesn't create any collisions between bound identifiers and free identifiers. If the identifier z is free in e, then we need to be sure that the β reduction doesn't make z become bound. If there is a name collision between a variable that is bound in B and a variable that is free in e, then we need to use α to change the identifier names so that they're different.

As usual, an example will make that clearer. Suppose we have a expression defining a function, $\lambda z . (\lambda x. x + z)$. Now suppose we want to apply it: $(\lambda z . (\lambda x . x + z)) (x + 2)$. In the parameter $(x + 2)$, x is free. Next, suppose we break the rule and go ahead and do β. We'd get $\lambda x . x + x + 2$. The variable that *was* free in $x + 2$ is now bound! We've changed the meaning of the function, which we shouldn't be able to do. If we were to apply that function after the incorrect β, we'd get $(\lambda x . x + x + 2)\ 3$. By β, we'd get $3 + 3 + 2$, or 8.

What if we did α the way we were supposed to?

First, we'd do an α to prevent the name overlap. By α[x/y], we would get $\lambda z . (\lambda y . y + z) (x+2)$.

Then by β, we'd get $\lambda y . y + x + 2$. If we apply this function the way we did before, then by β, we'd get $3 + x + 2$. $3 + x + 2$ and $3 + 3 + 2$ are very different results!

That's all that you can do in a λ calculus computation. All computation is really just β reduction, with α renaming to prevent name collisions. In my experience, that makes it the simplest formal system of computation. It's a lot simpler than the state-plus-tape notion of the Turing machine, which is itself one of the simpler ways of doing computation.

If that's too simple for you, there's another *optional* rule you can add called η ("eta") conversion. η is a rule that adds *extensionality*, which provides a way of expressing equality between functions.

eta says that in any λ expression, you can replace the value f with the value g as long as for all possible parameter values x, $f\,x = g\,x$.

Programming Languages and Lambda Strategies

In the beginning of this chapter, I said a lot about how useful λ calculus is for talking about programming-language design. From just looking at the calculus this far, exactly what λ calculus brings to the table isn't entirely clear. To get a sense of where the value comes from, we're going to take a quick look at λ calculus evaluation strategies and their relationship to programming languages.

In any course on programming languages, you're going to get a lecture about different evaluation strategies: eager evaluation versus lazy evaluation.

These ideas all involve the mechanics of parameter passing in the programming language. Suppose you're thinking about a call to a function `f(g(x+y), 2*x)`. In a language with *eager evaluation*, you'd first compute the values of the parameters and only invoke the function after the parameters had been computed. So in this example, you'd compute `g(x+y)` and `2*x` before invoking `f`; and when you were computing `g(x+y)`, you'd first compute `x+y` before invoking the function `g`. This is the way that many familiar languages actually work: for example, this is how C, Java, JavaScript, and Python work.

In *lazy evaluation*, you don't compute the value of any expression until you need to. In our example code, you'd invoke `f` first. You wouldn't invoke `g(x+y)` until `f` tried to use the value of that expression. If `f` never specifically used the value of the parameter expression `g(x+y)`, then it would never get computed and `g` would never get invoked. This kind of evaluation turns out to be really useful, and it's the basis of languages like Haskell and Miranda.

Defining exactly what eager versus lazy evaluation means, in a precise enough way to make things really predictable, is difficult. Unless you use λ calculus.

In λ calculus, as we've seen, computation is really done by repeated use of β reductions. If you look at a λ calculus expression, there are typically a lot of different β reductions that you can perform at any given moment. So far, we've been ad hoc about that, choosing which β reduction to perform when based on what was clearest for explaining the meaning of a construction. If you want to think about programming languages, you can't be ad hoc: a language needs to be predictable! You need to specify how you're going to do your β reductions in a precise and reproducible way.

The way that you perform your β reductions is called the *evaluation strategy*. The two most common evaluation strategies are these:

Applicative Order In applicative order, find the innermost expressions that are β reducible and perform them from right to left. Effectively, you're looking at your λ expression as a tree and evaluating the tree from right to left, from the leaves of the tree upward.

Normal Order In normal order, you start with the outermost expressions and evaluate them from left to right.

Applicative order is exactly what we meant by *eager evaluation,* and normal order is *lazy evaluation*. Let's look at an example to see the difference: `(λ x y z . + (* x x) y) (+ 3 2) (* 10 2) (/ 24 (* 2 3))`.

Applicative (Eager) Order In applicative order, we start with the innermost expressions and evaluate them from right to left. The innermost expression in this is the `(* 2 3)`. So we'd do the β reductions to evaluate it and reduce it to 6. Then right to left, we'd evaluate `(/ 24 6)`, `(* 10 2)`, `+ 3 2`. That would reduce our expression to `(λ x y z . + (* x x) y) 5 20 4`. Next, we'd reduce the outermost λ: `(+ (* 5 5) 30)`, which would evaluate to `(+ 25 30)`, and finally to 55.

Normal Order In normal order, we'd start with the left-outermost first. So we'd *first* do the outer β reduction, giving us `(+ (* (+ 3 2) (+ 3 2)) (* 10 2)`.

The important thing to notice about the two different scenarios is that in applicative order evaluation, we evaluated all of the parameters first; in normal order evaluation, we didn't evaluate the parameters until we needed to. In the case of the normal order evaluation, that meant that we never needed to evaluate the parameter expression `(/ 24 (* 2 3))` because we never used that value.

λ calculus shows us that the two evaluation strategies perform the same computation in the sense that they produce the same result. It also gives us an extremely simple way of defining what laziness means. We usually say that lazy evaluation means we don't evaluate any expression until we need to, but that doesn't explain how we know when we need to evaluate something. Normal order evaluation defines laziness for us: you need to evaluate an expression when it's the leftmost outermost unevaluated expression.

Similarly, we can show how different parameter-passing strategies like call-by-value, call-by-reference, and call-by-name all work by describing the evaluation strategy in λ calculus.

Now we've seen our first taste of λ calculus: how to read and write it and how to evaluate it. We've seen a bit of why λ calculus is useful by showing how different ways of ordering β reductions are used to describe different kinds of programming-language semantics.

We're still missing some really important things. We've handwaved using numbers, but we don't really know how to make them work: we know that in λ calculus, the only really computational step is β reduction, but we don't know how to do arithmetic using nothing but β reduction. Similarly, we don't know how to do conditionals or iteration using β reduction. Without numbers, iteration, and conditionals, λ calculus wouldn't be Turing complete!

In the next chapter, we'll take care of that by showing how to fill in all of those gaps.

25

Numbers, Booleans, and Recursion

But Is It Turing Complete?

λ calculus, as we described it in the previous chapter, is Turing complete. The question that you should be asking yourself now is how. Remember the three properties of computation? You need to have unbounded storage; you need to be able to do arithmetic; and you need to be able to do control flow. How can we do that in λ calculus?

The storage part is easy. You can store arbitrarily complicated values in a variable, and we can generate as many functions as we need without bound. So unbounded storage is pretty obvious.

But arithmetic? In our work so far we've just handwaved arithmetic by treating it as a primitive. In fact, we can create a very cool way of doing arithmetic using nothing but λ expressions.

And what about choices and repetition? So far we have no way of making choices between alternatives, or of repeating operations. It's hard to imagine how we're going to get there: at first glance, it seems like there's so little capability in λ expression evaluation, where all we can do is rename or replace things. We'll see that there's a way of making choices based on the way that we'll do arithmetic. And repetition in λ calculus is done using a really amazing trick. Repetition in λ calculus can only be done by using recursion and that can only be done by using something called a *fixed-point combinator*.

None of these things that we need to be Turing complete are built into λ calculus. But fortunately, it's all stuff that we can build. So in this chapter we'll look at how to build the things that we need in λ calculus.

Before we move on, for convenience, I'll introduce a way of naming things. In the programming-language world, we call that *syntactic sugar*—it's just a notational shorthand—but when we start looking at some more complicated expressions, it makes a huge difference in readability.

We'll define a *global* function (that is, a function that we'll use throughout our λ calculus introduction without including its declaration in every expression) like this:

$square = \lambda\, x\, .\, x * x$

This declares a function named *square*, whose definition is $\lambda\, x\, .\, x * x$. If we had an expression *square 4*, the definition we gave means that it would effectively be treated as if the expression were *(λ square . square 4)(λ x . x×x)*.

Numbers That Compute Themselves

To show that λ calculus is Turing complete, we said we need to show two more things. We need to be able to do arithmetic, and we need to be able to do flow control. For arithmetic, we get—once again!—to create numbers. But this time we'll do them with λ expressions. We'll also see how to take the same basic mechanics that we'll use to create numbers, and turn them into a form of conditional if/then/else construct, which will give us the first half of what we need for full flow control in λ calculus.

As we've seen, all we have to work with in λ calculus is functions written as λ expressions. If we want to create numbers, we have to do it by devising some way of creating objects that we can use to do Peano arithmetic using nothing but functions. Fortunately for us, Alonzo Church, the genius who invented λ calculus, worked out how to do that. His version of numbers-as-functions are called *Church numerals*.

In Church numerals, all numbers are functions with two parameters: *s* (for *successor*) and *z* (for *zero*).

- *Zero* = $\lambda\ s\ z\ .\ z$
- *One* = $\lambda\ s\ z\ .\ s\ z$
- *Two* = $\lambda\ s\ z\ .\ s\ (s\ z)$
- *Three* = $\lambda\ s\ z\ .\ s\ (s\ (s\ z))$
- And so on. Any natural number *n* is represented by a Church numeral, a function that applies its first parameter to its second parameter *n* times.

A good way of understanding this is to think of *z* as being a name for a function that returns a zero value and *s* as a name for a successor function.

Church numerals are absolutely amazing. Like so much else in λ calculus, they're a tribute to the astonishing brilliance of Alonzo Church. The beauty of them is that they're not just representations of numbers: they're a direct representation of the computation to generate the numbers they represent using the Peano axioms. Here's what I mean: imagine that we had another way of representing numbers. Then we could write a zero function and a successor function in that new representation. For example, we could implement unary numbers using strings:

- *UnaryZero* = $\lambda\ x.$ *""*
- *UnarySucc* = $\lambda\ x.$ *append "1" x*

If you took the Church numeral for the number 7, $\lambda\ s\ z\ .\ s(s(s(s(s(s(s(z)))))))$, and applied it to *UnaryZero* and *UnarySucc*, then the result would be 1111111, the unary representation of 7.

Addition works off of the same self-computation principle. If we have two numbers *x* and *y* that we want to add, we can invoke *x* using *y* as the zero function, and *x* would *add itself* to *y*.

Actually, it's a bit more complicated than that because we need to be sure that *x* and *y* use the same increment function. If we want to do addition, $x + y$, we need to write a function with four parameters; the two numbers to add and the *s* and *z* values we want in the resulting number:

add $\lambda\ s\ z\ x\ y\ .\ x\ s\ (y\ s\ z)$

There are two things going on in that definition: First, it's taking two parameters that are the two values we need to add; second, it needs to normalize things so that the two values being added end up sharing the same binding of the zero and successor values. To see how those two pieces of addition work, let's curry the definition in order to separate them out.

add_curry = λ *x y*. (λ *s z* . (*x s* (*y s z*)))

Looking at that carefully, *add_curry* says that if you want to add *x* and *y*, you need to do this:

1. Create the Church numeral *y* using the parameters *s* and *z*.

2. Take the result and apply *x* to it, using the same *s* and *z* functions.

As an example, let's use *add_curry* to add the numbers 2 and 3.

Example: Adding 2 + 3 using a curried function

1. *two* = λ *s z* . *s* (*s z*)

2. *three* = λ *s z* . *s* (*s* (*s z*))

3. Now we want to evaluate: *add_curry* (λ *s z* . *s* (*s z*)) (λ *s z* . *s* (*s* (*s z*)))

4. Using the same names in *two* and *three* is going to be a problem, so we'll α them and make *two* use s2 and z2 and *three* use s3 and z3; that gives us this: *add_curry* (λ *s2 z2* . *s2* (*s2 z2*)) (λ *s3 z3* . *s3* (*s3* (*s3 z3*)))

5. Now let's replace "add_curry" with its definition: (λ *x y* .(λ *s z*. (*x s y s z*))) (λ *s2 z2* . *s2* (*s2 z2*)) (λ *s3 z3* . *s3* (*s3* (*s3 z3*)))

6. Do a β on the outermost function application: λ *s z* . (λ *s2 z2* . *s2* (*s2 z2*)) *s* (λ *s3 z3* . *s3* (*s3* (*s3 z3*)) *s z*)

7. Now we get to the interesting part: we're going to β the Church numeral for *three* by applying it to the *s* and *z* parameters. This normalizes *three*: it replaces the successor and zero function in the definition of *three* with the successor and zero functions from the parameters to add. This is the result: λ *s z* . (λ *s2 z2* . *s2* (*s2 z2*)) *s* (*s* (*s* (*s z*)))

8. And we β again, this time on the λ for *two*. Look at what we're going to be doing here: *two* is a function that takes two

parameters: a successor function and a zero function. To add *two* and *three*, we're using the successor function that was passed to the outer *add_curry* function, and we're using the result of evaluating *three* as the value of the zero for *two*: $\lambda\, s\, z\, .\, s\, (s\, (s\, (s\, (s\, z))))$

9. And we have our result: the Church numeral for five!

Aside from being pretty much the coolest representation of numbers in the known universe, Church numerals also set the pattern for how you build computations in λ calculus. You write functions that combine other functions in order to make it do what you want.

Decisions? Back to Church

With numbers out of the way, we're closing in on Turing completeness. We're still missing two things: the ability to make decisions, and repetition.

To make decisions, we're going to do something very similar to how we did numbers. To represent numbers, we built functions that computed the numbers. To do choice, we're going to do almost the same trick: we're going to create Boolean values that select alternatives.

For making decisions, we'd like to be able to write choices as if/then/else expressions, like we have in most programming languages. Following the basic pattern of the Church numerals, where a number is expressed as a function that adds itself to another number, we'll express true and false values as functions that perform an if/then/else operation on their parameters. These are sometimes called *Church Booleans* (of course, they were also invented by Alonzo Church). An if/then/else choice construct is based on the two Boolean values, true and false. In λ calculus, we represent them as (what else?) functions. They're two-parameter functions that take two arguments:

- $TRUE = \lambda\, t\, f\, .\, t$
- $FALSE = \lambda\, t\, f\, .\, f$

With the Church Booleans, it's downright simple to write an *if* function whose first parameter is a condition expression, whose second parameter is the expression to evaluate if the condition is true, and whose third parameter is the expression to evaluate if the condition is false.

IfThenElse = λ *cond t f* . *cond t f*

We can also build the usual Boolean operations:

- *And* = λ *x y* . *x y FALSE*.
- *BoolOr* = λ *x y*. *x TRUE y*.
- *BoolNot* = λ *x* . *x FALSE TRUE**.

Let's take a closer look at those to see how they work. Let's first take a look at *BoolAnd*.

Let's start by evaluating *BoolAnd TRUE FALSE*.

1. Expand the *TRUE* and *FALSE* definitions: *BoolAnd (λ t f . t) (λ t f . f)*.

2. α the true and false: *BoolAnd (λ tt tf . tt) (λ ft ff . ff)*.

3. Now expand BoolAnd: *(λ t f. t f FALSE) (λ tt tf . tt) (λ ft ff . ff)*.

4. β: *(λ tt tf.tt) (λ ft ff. ff) FALSE*.

5. β again: *(λ xf yf . yf)*.

And we have the result: *BoolAnd TRUE FALSE* = *FALSE*. Now let's try *BoolAnd FALSE TRUE*:

1. *BoolAnd (λ t f . f) (λ t f .t)*.

2. α: *BoolAnd (λ ft ff . ff) (λ tt tf . tt)*.

3. Expand BoolAnd: *(λ x y .x y FALSE) (lambda ft ff . ff) (lambda tt tf . tt)*.

4. β: *(λ ft ff . ff) (lambda tt tf . tt) FALSE*.

5. β again, and you end up with *FALSE*. So *BoolAnd FALSE TRUE* = *FALSE*.

Finally, let's try *BoolAnd TRUE TRUE*:

1. *BoolAnd TRUE TRUE*

2. Expand the two trues: *BoolAnd (λ t f . t) (λ t f . t)*.

3. α: *BoolAnd (λ xt xf . xt) (λ yt yf . yt)*.

4. Expand BoolAnd: *(λ x y . x y FALSE) (λ xt xf . xt) (λ yt yf . yt)*.

5. β: *(λ xt xf . xt) (λ yt yf . yt) FALSE.*
6. β again: *(λ yt yf . yt).*
7. So *BoolAnd TRUE TRUE = TRUE.*

The other Boolean operations work in much the same way. By the genius of Alonzo Church, we have *almost* everything we need to show that λ calculus is Turing complete. All that's left is recursion. But recursion in λ calculus is a real mind-bender!

Recursion: Y Oh Y Oh Y?

We've been building up the bits and pieces that turn λ calculus into a useful system. We've got numbers, Booleans, and choice operators. The only thing we're lacking is some kind of repetition or iteration.

In λ calculus, all iteration is done by recursion. In fact, recursion is a pretty natural way of expressing iteration. It takes a bit of getting used to, but if you spend a lot of time in a functional language like Scheme, ML, or Haskell, you get used to it; and then when you return to an imperative language like Java, there's a good chance that you'll end up feeling frustrated about having to force all of your iterations into the loop structure instead of just being able to do them recursively!

It can be a bit difficult if you're not used to thinking recursively. So we'll start by looking at the basics of recursion.

Understanding Recursion

The cleverest definition that I've seen of recursion comes from *The New Hacker's Dictionary [Ray96]*, whose entry reads as follows:

> Recursion: see "Recursion."

Recursion is about defining things in terms of themselves. It almost seems like magic until you get the hang of it. The principle is the same as what we saw with induction back in 1, *Natural Numbers*, on page 3, but applied to a definition instead of a proof.

It's easiest to show you what that means with an example, like a factorial. Let's think about factorials. The factorial function *N!* is defined for all natural numbers: for any natural number *N*, its factorial is the product of all of the integers less than or equal to *N*.

- 1! = 1
- 2! = 1 * 2 = 1
- 3! = 1 * 2 * 3 = 6
- 4! = 1 * 2 * 3 * 4 = 24
- 5! = 1 * 2 * 3 * 4 * 5 = 120
- And so on

If you look at that definition, it's pretty cumbersome. If you you look at the sequence of examples, you can see that there's a pattern: the factorial of each number *N* is the product of a sequence of numbers, and that sequence is exactly the same as the sequence for the number before it, with *N* added onto the end.

We can use that to make the list a bit simpler; let's just replace the sequence for everything but *N* with the product of the numbers in that part of the sequence:

- 1! = 1
- 2! = 1 * 2 = 2
- 3! = 2 * 3 = 6
- 4! = 6 * 4 = 24
- 5! = 24 * 5 = 120
- And so on

Now the pattern should be reasonably easy to see: look at 4! and 5!; 4! = 24; 5! = 5 * 24. Since 4! is 24, we can then say that 5! = 5 * 4!.

In fact, we can say in general that for any *N*, *N! = N * (N-1)!*.

Well, almost.

The expression doesn't quite work, because it *never stops*. Try computing *3!*: *3! = 3 * 2! = 3 * 2 * 1! = 3 * 2 * 1 * 0! = 3 * 2 * 1 * 0 * -1! = ...*

We could keep going forever, because the way that we defined it, we never stop repeating. The definition doesn't have any way to stop.

To make recursion work, you need to give it a way to eventually *stop*. In formal terms, you need to define a *base case*: that is, you need to define a place where the recursion stops—where you've got a definition for *some value* that can be computed without doing any more recursion.

For factorial, we do that by saying that the factorial of 0 is 1. Then things work: factorial is only supposed to work for positive numbers; and for any positive number, the recursive definition will expand until it hits 0!, and then it will stop. So let's look at 3! again: $3! = 3 * 2! = 3 * 2 * 1! = 3 * 2 * 1 * 0! = 3 * 2 * 1 * 1 = 6$.

What we saw in the factorial is what we'll see in every recursive definition. The definition will be written with two cases: a general case that defines the function recursively and a specific base case that defines the function for specific values nonrecursively.

The way that we write a recursive definition is to state the two cases using conditions in something that looks almost like a little bit of a program:

- $N! = 1$ if $N = 0$
- $N! = N * (N-1)!$ if $N > 0$

Congratulations! Now you understand at least a little bit of recursion.

Recursion in λ Calculus

Suppose we want to write a factorial in λ calculus. We're going to need a few tools. We need a test for equality to zero, we need a way of multiplying numbers, and we need a way of subtracting 1.

For testing equality to zero, we'll use a function named *IsZero*, which takes three parameters: a number and two values. If the number is zero, it returns the first value; if it's not zero, then it returns the second value.

Multiplication is an iterative algorithm, so we can't write multiplication until we work out recursion. But we'll just handwave that for now and have a function *Mult x y*.

And finally, for subtracting 1, we'll use *Pred x* for the predecessor of *x*—that is, *x - 1*. So a first stab at a factorial written with the recursive call left with a blank in it would be this:

*λ n . IsZero n 1 (Mult n (**something** (Pred n)))*

Now the question is, what kind of *something* can we plug in there? What we'd really like to do is plug in a copy of the function itself:

Fact = λ n . IsZero n 1 (Mult n (Fact (Pred n)))

How can we do that? The usual way of plugging something into a λ calculus function is by adding a parameter:

Fact = (λ f n . IsZero n 1 (Mult n (f (Pred n)))) Fact

We can't plug in a copy of the function as its own parameter that way: the name *Fact* doesn't exist in the expression in which we're trying to use it. You can't use an undefined name, and in λ calculus, the only way to bind a name is by passing it as a parameter to a λ expression. So what can we do?

The answer is to use something called a *combinator*. A combinator is a special kind of function that operates on functions and that can be defined without reference to anything but function applications. We're going to define a special, almost magical function that makes recursion possible in λ calculus, called the *Y combinator*.

Y = λ y . (λ x . y (x x)) (λ x . y (x x))

The reason for calling it *Y* is because it is shaped like a *Y*. To show you that more clearly, sometimes we write λ calculus using trees. The tree for the *Y* combinator is in Figure 18, *The Y Combinator*, on page 241.

Why is the *Y* combinator an answer to our problem in defining the factorial function? The *Y* combinator is a *fixed-point combinator*. That means that it's a strange beast that's capable of reproducing itself! What makes it special is the fact that for any function *f*, *Y f* evaluates to *f Y f*, which evaluates to *f (f Y f)*, which evaluates to *f (f (f Y f))*. See why it's called *Y*?

Let's try walking through *Y f*:

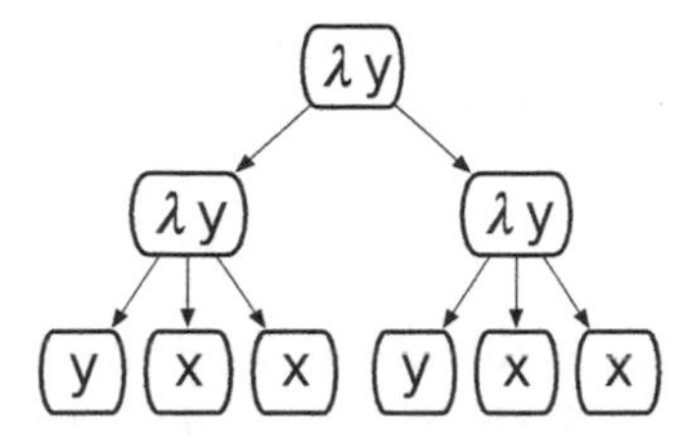

Figure 18—The *Y* Combinator: *When the Y combinator is drawn as a tree, it's clear where the "Y" in its name originates.*

- Expand *Y*: $(\lambda\, y\, .\, (\lambda\, x\, .\, y\, (x\, x))\, (\lambda\, x\, .\, y\, (x\, x)))\, f$.
- β: $(\lambda\, x\, .\, f\, (x\, x))\, (\lambda\, x\, .\, f\, (x\, x))$.
- β again: $f\, (\lambda\, x\, .\, f\, (x\, x))\, (\lambda\, x\, .\, f\, (x\, x))$.
- Since $Y\, f = (\lambda\, x\, .\, f\, (x\, x))\, (\lambda\, x\, .\, f\, (x\, x))$, what we just got in step three is $f\, Y\, f$.

See, there's the magic of *Y*. No matter what you do, you can't make it consume itself. Evaluating *Y f* will produce another copy of *f* and leave the *Y f* part untouched.

So how do we use this crazy thing?

Remember our last attempt at defining the factorial function? Let's look at it again:

Fact = $(\lambda\, f\, n\, .\, IsZero\ n\ 1\ (Mult\ n\ (f\ (Pred\ n))))$ *Fact*

Let's rewrite it just a tiny bit to make it easier to talk about:

Metafact = $(\lambda\, f\, n\, .\, IsZero\ n\ 1\ (Mult\ n\ (f\ (Pred\ n))))$

With that, *Fact* = *Metafact Fact*.

Now, we're left with one last problem. *Fact* is not an identifier defined inside of *Fact*. How do we let *Fact* reference *Fact*? Well, we did a λ abstraction to let us pass the *Fact* function as a parameter, so what we need to do is to find a way to write *Fact* that lets us pass it to itself as a parameter.

But remember what *Y f* does? It expands into a call to *f* with *Y f* as its first parameter. In other words, *Y f* turns *f* into a recursive function with *itself* as its first parameter! So the factorial function is this:

Fact = *Y Metafact*

I learned about the *Y* combinator back in my undergrad days, which would place that around 1989, and I still find it rather mystifying. I understand it now, but I can't imagine how on earth anyone ever figured it out!

If you're interested in this, then I *highly* recommend you get yourself a copy of the book *The Little Schemer [FFB95]*. It's a wonderful little book. It's set up like a children's book, where the front of each page is a question and the back of each page is the answer. It's written in a delightfully playful style, it's very fun and engaging, and it will also teach you to program in Scheme.

We've seen everything that we need to write arbitrary computations in λ calculus. We've shown that it gives us arbitrary amounts of storage using variables and complex values. We've seen how to build numbers using the amazing self-computation trick of Church numerals. We figured out how to do choice using Church Booleans. And finally, we saw how to do repetition using recursion with the *Y* combinator. It took some effort, but with the tools we've built, λ calculus can do anything we want it to do.

The power of this is remarkable, and as a result, it's been used all over. Most notably to people like me, there's probably no programming language in use anywhere that hasn't been influenced at least a little bit by λ calculus.

Unfortunately, we still have a problem. Just like everything else in math, a calculus like λ needs a *model*. The model shows that the calculus, the way that we defined it, is really valid. Without a model, λ calculus could be fooling us: it could look like something terrific, but like we saw with Russell's paradox in naive set theory, it could be built on a fundamental flaw that makes it inconsistent!

In the next chapter, we'll look at how to fill that hole and show that there is a valid model for λ calculus. Along the way, we'll see just what *types* are and how types can help detect errors in a program using logic.

26

Types, Types, Types: Modeling λ Calculus

λ calculus began with the simple untyped λ calculus that we discussed in the previous chapter. But one of the great open questions about λ calculus at the time of its invention was this: is it sound? In other words, does it have a valid model?

Church was convinced that λ calculus was sound, but he struggled to find a model for it. During his search, Church found that it was easy to produce some strange and worrisome expressions using the simple λ calculus. In particular, he was worried about falling into a Gödel-esque trap of self-reference (which we'll talk about more in 27, *The Halting Problem*, on page 253), and he wanted to prevent that kind of inconsistency. So he tried to distinguish between values representing atomic primitive values and values representing predicates. By making that distinction, he wanted to ensure that predicates could operate only on atoms and not on other predicates.

Church did this by introducing the concept of *types*. Types provided a way of constraining expressions in the calculus that made it impossible to form the kinds of self-referential structures that could lead to inconsistencies. The addition of types created a new version of λ calculus that we call the *simply typed λ calculus*. The original goal of this was to show that λ calculus had a valid model. The idea that it introduced turned out to be useful for far more than that: if you've ever used a statically typed programming language, you've seen the products of λ calculus: the typed λ calculus first introduced by Alonzo Church is the foundation of all of the type systems and of the very notion of types that we use.

The first version of typed λ calculus is called simply typed because it's the simplest reasonable notion of types. It provides base types and function types, but not types with parameters or predicates. In programming language terms, using this version of λ calculus, we could define a type *Integer*, but we couldn't define a parametric type like *List of Integer*.

Playing to Type

When Church designed typed λ calculus, his goal was to build a model that showed that λ calculus was consistent. What he was worried about was a Cantor-esque self-reference problem. In order to avoid that, he created a way of partitioning values into groups called *types* and then used that idea of types to constrain the language of λ calculus so that you couldn't write an expression that did something inconsistent.

The main thing that typed λ calculus adds to the mix is a concept called base types. In a typed λ calculus, you have some universe of atomic values that you can manipulate. Those values are partitioned into a collection of distinct non-overlapping groups called the *base types*. Base types are usually named by single lowercase Greek letters. In our description of the simply typed λ calculus, we'll use α for a type containing the natural numbers, β for a type containing Boolean true/false values, and γ ("gamma") for a type containing strings.

Once we have basic types, we can then talk about the type of a function. A function maps from a value of one type (the type of its parameter) to a value of a second type (the type of the return value). For a function that takes a parameter of type γ and returns a value of type δ ("delta"), we write its type as $\gamma \rightarrow \delta$. The right-arrow is called the *function type constructor*; it associates to the right, so $\gamma \rightarrow \delta \rightarrow \varepsilon$ ("epsilon") is equivalent to $\gamma \rightarrow (\delta \rightarrow \varepsilon)$. The resemblance to a logical implication isn't accidental: a function type $\alpha \rightarrow \delta$ *is* a logical implication: passing a value of type α to the function implies that the return type will be β.

To use these types in λ calculus, we need to add two new constructs. First, we need to modify our λ calculus syntax so that we can include type information in λ terms. Second, we need to add a set of rules to show what it means for a typed program to be valid.

The syntax part is easy: we add a colon (":") to the notation. The colon binds the expression or variable binding on its left to the type specification on its right. It asserts that whatever is on the left side of the colon has the type specified on the right side. This method of specifying the type of an expression is called a *type ascription.*

Let's look at a few examples:

- $\lambda x{:}\ \alpha\ .\ x + 3$: This is a simple function that declares that the parameter x has type α, which is our name for the natural numbers. This function doesn't say anything about the type of the result of the function, but since we know that + is a function with type $\alpha \rightarrow \alpha$, we can infer that the result type of this function will be α.

- $(\lambda\ x\ .\ x + 3){:}\ \alpha \rightarrow \alpha$: This is the same as the previous function with a different but equivalent type declaration. This time the type ascription asserts the type of the entire λ expression. We can infer that $x : \alpha$ because the function type is declared to be $\alpha \rightarrow \alpha$, which means that the parameter has type α.

- $\lambda x : \alpha, y : \delta\ .\ if\, y\ then\ x * x\ else\, x$. Now we're getting more complicated. This is a two-parameter function where the first parameter is type α and the second parameter is type δ. We can infer the return type, because it's the type of x that is α. Using that, we can see that the type of the full function is $\alpha \rightarrow \delta \rightarrow \alpha$. This may seem surprising at first because it's a two-parameter function, but we're writing the type using multiple arrows. The trick here is that, as I explained in Section 24.1, *Writing λ Calculus: It's Almost Programming!*, on page 220, λ calculus really works entirely in terms of single-parameter functions; multiparameter functions are a shorthand for currying. $x{:}\ \alpha\, y{:}\delta\ .\ if\, y\ then\ x * x\ else\, x$ is shorthand for λ

$x : \alpha . (\lambda\, y : \delta\, .\, if\, y\ then\ x * x\ else\ x)$. The inner λ has type $\delta \rightarrow \alpha$, and the outer λ has type $\alpha \rightarrow (\delta \rightarrow \alpha)$.

The point of typing is to enforce consistency on λ calculus expressions and programs. To do that, we need to be able to determine whether or not a program's use of types is consistent and valid. If it is, we say that the program is *well typed*. The way that we check whether a program is well typed is by using a system of *type inference*. Type inference takes type declarations as axioms and uses logical inference to determine the types of every clause and expression in the program. If the type of every expression can be inferred and none of the inferred types is different from the declared types, then the program is well typed. When the type of an expression is inferred using the type logic, we call that inference a *type judgement*.

Type judgements are usually written in a notation called a *sequent*, which looks like a fraction where the numerator consists of statements that we know to be true and the denominator is what we can infer from the numerator. In the numerator, we normally have statements using a *type context* (or just *context*), which is a set of type judgements that we already know. The context is usually written as an uppercase letter. If a type context G includes the judgement that $x : \gamma$, we'll write that as G :- $x : \gamma$".

For the simply typed λ calculus, a simplified version of the type inference rules are given here:

Type Identity

$$\frac{}{x : \alpha \models x : \alpha}$$

This is the simplest rule: if we have no information other than a type ascription of the variable's type, then we know that that variable has the type that was ascribed to it.

Type Invariance

$$\frac{G \models x : \alpha,\ x \neq y}{G + y : \beta \models x : \alpha}$$

This is a statement of noninterference. It says that once we've been able to judge the type of some variable or expression, adding more information about the types of other variables or expressions can't change our judgement. This is extremely important for keeping type inference usable in practice; it means that we can make a judgement as soon as we have enough information, and we never need to reevaluate it.

Function Type Inference

$$\frac{G + x : \alpha \models y : \beta}{G \models (\lambda x : \alpha . y) : \alpha \rightarrow \beta}$$

This statement allows us to infer function types given parameter types. If we know the type of the parameter to a function is α, and we know that the type of the term that makes up the body of the function is β, then we know that the type of the function is $\alpha \rightarrow \beta$.

Function Application Inference

$$\frac{G \models x : \alpha \rightarrow \beta,\ G \models y : \alpha}{G \models (xy) : \beta}$$

If we know that a function has type $\alpha \rightarrow \beta$ and we apply it to a value of type α, the result is a value of type β.

These four rules are all that we need. If we can come up with a consistent set of type judgements for every term in a λ expression, then the expression is well typed. If not, then the expression is invalid.

Let's work through an example.

*Example: λ x y . if y then 3 * x else 4 * x.*

1. If/then/else would be a built-in function, which would be in the context. The type would be $\beta \rightarrow a \rightarrow a \rightarrow a$—that is, it's a function that takes three parameters: a Boolean, a value to return if the Boolean is true, and a value to return if the Boolean is false. For if/then/else to be well typed, both the second and third parameters must have the same type or the function would have two possible return types, which would be inconsistent. So using the known type information about if/then/else and using function type inference, we can infer that y must have the type β.
2. Similarly, we do the same basic thing with the other expressions in the if/then/else function. * is a function from a number to a number, so since we're using x as a parameter to *, $x : \alpha$.

3. Once we know the types of the parameters, we can use application inference to find the return type of *if/then/else*: it's $\beta \to \alpha \to \alpha \to \alpha$.

4. And finally, we can use function type inference now to get the type of the top-level function. *(λ x: α y: β . if y then 3 * x else 4 * x): α → β → α.*

In more complex programs, we often need to play with *type variables*. A type parameter is a placeholder for a currently unknown type. When we have a variable whose type we don't know, we can introduce a *new* type variable. Eventually, when we figure out what the actual type should be in any place where the type variable is used, we can replace it. For example, let's look at a really simple expression: $\lambda\, x\, y.y\, x$.

Without any type declarations or parameters, we don't know its exact type. But we do know that *x* has some type, so we'll use a *type variable* to represent its unknown type and hope that later we'll be able to replace the variable with a specific type. We'll call the type variable *t*, which means that using type identity, we can add the judgement *x: t*. We know that *y* is a function because it's applied in the λ body and takes *x* as a parameter. Since we don't know what type it returns, we'll use another new type variable *u*, and say *y: u* (function type inference). By doing function application inference, we can judge that the result type of the application of *y* is *u*. This means that we can write types for everything in that function using type variables: *(λ x:t y: t → u. (y x): u) t → (t → u) → u*. We can't infer anything else without knowing more about the types of *x* and *y*. So we've been able to infer a lot just from the structure of the λ expression, but we can't quite get to the point of showing that it's well typed.

To see the problem, suppose that we wanted to apply our λ like this: *(λ x y . y x) 3 (λ a . if a then 3 else 2)*. Then we'd be able to say that *y* must have type $\beta \to \alpha$ (we said earlier that β is the type of Boolean values, and α is natural numbers). Since we're passing 3 for *x*, then *x: α*. Now we've got an inconsistency: according to the type judgement for *y*, *t* must be *β*, but according to the type judgement for *x*, *t* must be *α*.

That's all we needed: now we have the simply typed λ calculus.

Let's take another look at the types of the simply typed λ calculus. Anything that can be formed from the following grammar is a λ calculus type:

Prove It!

type : : = *primitive* || *function* || (*type*)
primitive : : = *alpha;* || *beta;* || *delta;* || ···
function : : = *type* *rarr; type*

The grammar defines the syntax of valid types, but that's not quite enough to make the types meaningful. Using that grammar, you can create type expressions that are valid but for which you can't actually write an expression that will produce a value of that type. When there is an expression that has a type, we say that the expression *inhabits* the type and that the type is an inhabited type. If there is no expression that can inhabit a type, we say it's *uninhabitable*. So what's the difference between an inhabitable type and an uninhabitable type?

The answer comes from something called the *Curry-Howard isomorphism*. The Curry-Howard isomorphism is one of the most brilliant pieces of work that I've ever seen. It showed that for a typed λ calculus, there is a corresponding intuitionistic logic: a type expression in the λ calculus is inhabitable if and only if the type is a *provable theorem* in the corresponding logic.

I alluded to this earlier: look at the type $\alpha \rightarrow \alpha$. Instead of seeing "→" as the function type constructor, try viewing it as a logical implication. "α implies α" is clearly a theorem of intuitionistic logic. So the type $\alpha \rightarrow \alpha$ is inhabitable.

Now look at $\alpha \rightarrow \beta$. That's not a theorem, unless there's some other context that proves it. As a function type, we can understand it as the type of function that takes a parameter of type α and returns something different, a value of type β. Taken on its own, you can't do that: the β needs to come from *somewhere*. In order for $\alpha \rightarrow \beta$ to be a theorem, it must be provable. What's the proof? A program that takes a parameter of type α and returns a value of type β. The program *is the proof* that the type is inhabitable.

The fact that λ calculus programs are proofs is even deeper than that. You can take any statement in intuitionistic logic and render it into a type declaration in λ calculus. Then you can *prove* the validity of that statement by writing a valid program. The program *is* the proof; the β reductions that are used to evaluate the proof are inference steps in the logic. There is a one-to-one relation between β reductions and inferences. And the execution of the program produces a concrete example that demonstrates the truth of the statement.

When we began this chapter, I said that in order for it to be considered sound, λ calculus needed a valid model. And here we are: intuitionist logic is the model.

What's It Good For?

Aside from making it possible to build a model, types made it possible to reason about expressions in the λ calculus in amazing ways. Types for λ calculus changed the field of computation forever: not only are they useful for abstract mathematical study, but the typed λ calculus has directly impacted practical computation. Today it is widely used in the design of programming languages, as a tool for describing the meaning of programming languages, and even as a tool for describing the meaning of human languages. Type systems for λ calculi have never stopped developing: people are still finding new things to do by extending the λ type system today!

Most programming languages based on λ calculus are based on a variant of *System-F*, which extends λ calculus with a much more sophisticated type system that includes parameterized types. (If you want to learn more about System-F, a good introduction is *Types and Programming Languages [Pie02]*.) System-F was simplified and used in the design of a programming language called ML by Robin Milner (see *The Definition of Standard ML (Revised) [MHMT97]* for details), which was the basis for pretty much all of the modern typed λ-calculus-based programming languages. Milner went on to earn the Turing award for his work in designing ML as well as for his work in modeling concurrent computation using something called *the calculus of communicating systems*.

Once people really started to understand types, they realized that the untyped λ calculus was really just a pathologically simple instance of the simply typed λ calculus: a typed λ calculus with only one base type. Types weren't really necessary for the model, but they made it easier and opened new vistas for the practical application of λ calculus.

The simply typed λ calculus started as a way of constraining λ calculus expressions to ensure that they were consistent. But the way that Church created types is amazing. He didn't just create a constraint system. He added a level of *logic* to types. This was a stroke of absolute genius that meant that if a λ calculus function is well formed, then the types of its expressions will form a logical proof of its consistency! The type system of a simply typed λ calculus is an intuitionistic logic: each type in the program is a proposition in the logic, each β reduction corresponds to an inference step, and each complete function is a proof that the function contains no type errors!

27

The Halting Problem

What better way to finish a book on computing than by asking if computations will ever finish?

One of the most fundamental questions in computing is, can you tell if a particular computation will ever stop? It's obviously important to an engineer writing a program. A correct program should always eventually finish what it's doing and stop. If you've written a program that doesn't ever finish what it's doing, it's almost certainly incorrect.

There's a lot more to this question than just whether or not a particular program is correct. It gets to the very heart of questions about the fundamental limits of mathematics.

Most of the time when mathematicians talk about the limits of math, they end up talking about something called *Gödel's incompleteness theorem*. Incompleteness proved that not all true statements are provable. By showing this, the proof of incompleteness showed that the most ambitious project in the history of mathematics was doomed to be an utter failure. This made incompleteness simultaneously one of the greatest results and one of the greatest disappointments ever in all of math.

I'm not going to explain it.

What I am going to do is explain something related but simpler that comes from the question about whether a program will ever finish running and stop. We saw in 14, *Programming with Logic,* on page 103, that logical proofs and computations are the same thing. By looking at computations and asking whether or not those computations will ever finish, we can produce a result that means almost the same thing as incompleteness. Here's the fundamental question we're going to look at: if I give you a program *P*, can you write a program that will tell you whether *P* will ever finish running? Turing spent a lot of time working on it under its German name, the *Entscheidungsproblem*. In English, it's called the halting problem. But before I go into detail, I want to set the stage for why it's such a big deal!

A Brilliant Failure

Early in the twentieth century, a group of mathematicians led by Bertrand Russell and Alfred North Whitehead set out to do something amazing; they wanted to build a complete formalization of mathematics. They started with nothing but the most basic axioms of set theory and then tried to build up the complete edifice of mathematics as one system, published as the *Principia Mathematica*. The system described in *Principia* would have been the ultimate perfect mathematics! In this system, every possible statement would be either true or false: every true statement would be provably true, and every false statement would be provably false.

This should give you a sense of just how complex the Principia system was: it took Whitehead and Russell 378 pages of nothing but pure mathematical notation just to get to the point where they could prove that $1 + 1 = 2$. A brief except from that amazingly complicated proof is shown in Figure 19, *The Principia Mathematica*, on page 255.

If the Principia had been successful, it would have unlocked all of the secrets of mathematics. Russell and Whitehead would have accomplished what would have been the most significant intellectual accomplishment of all time.

Sadly, it was a failure. The system blew up in their faces: not only did their effort fail, but it was proven to be absolutely, utterly, inescapably impossible for it ever to succeed.

What happened to cause this magnificent effort to fall apart? Simple: Kurt Gödel (1906–1978) came along in 1931 and published a paper that was called "On Formally Undecidable Propositions in Principia Mathematica and Related Systems I," which contained his first *incompleteness theorem*.

Similarly $\vdash :. \beta \subset \iota`x \cup \iota`y \,.\, x \sim\in \beta \,.\, \supset : \beta = \Lambda \,.\, \vee \,.\, \beta = \iota`y$ (3)

$\vdash \,.\, (2) \,.\, (3) \,.\, {*}3{\cdot}48 \,.\, \supset$

$\vdash :. \beta \subset \iota`x \cup \iota`y \,.\, \sim(x, y \in \beta) \,.\, \supset : \beta = \Lambda \,.\, \vee \,.\, \beta = \iota`x \,.\, \vee \,.\, \beta = \iota`y$ (4)

$\vdash \,.\, (1) \,.\, (4) \,.\, {*}34{\cdot}8 \,.\, \supset$

$\vdash :. \beta \subset \iota`x \cup \iota`y \,.\, \supset : \beta = \Lambda \,.\, \vee \,.\, \beta = \iota`x \,.\, \vee \,.\, \beta = \iota`y \,.\, \vee \,.\, \beta = \iota`x \cup \iota`y$ (5)

$\vdash \,.\, {*}24{\cdot}12 \,.\, {*}22{\cdot}58{\cdot}42 \,.\, \supset$

$\vdash :. \beta = \Lambda \,.\, \vee \,.\, \beta = \iota`x \,.\, \vee \,.\, \beta = \iota`y \,.\, \vee \,.\, \beta = \iota`x \cup \iota`y : \supset \,.\, \beta \subset \iota`x \cup \iota`y$ (6)

$\vdash \,.\, (5) \,.\, (6) \,.\, \supset \vdash \,.\,$ Prop

This proposition shows that a class contained in a couple is either the null-class or a unit class or the couple itself, whence it will follow that 0 and 1 are the only numbers which are less than 2.

Figure 19—The *Principia Mathematica*: *An excerpt from the formal proof in* Principia Mathematica *(see pg. 378) that 1 + 1 = 2.* (Image courtesy of the University of Michigan Library archive at http://quod.lib.umich.edu/cgi/t/text/text-idx?c=umhist-math;idno=AAT3201.0001.001.)

Incompleteness proved that any formal system powerful enough to express Peano arithmetic would be either incomplete or inconsistent. The mechanics of the incompleteness proof are complex but stunningly beautiful.

Incompleteness absolutely halted work on any system like the Principia, because it showed that such efforts were doomed to fail. The theorem did this by showing that any complete system had to be inconsistent, and any consistent system had to be incomplete. What does that mean?

In mathematical terms, an inconsistent system is a system in which you can produce a proof of a false statement. In the math world, that's the worst possible flaw. In a logical system, if you can ever prove a false statement, that means that *every* proof in the system is invalid! We absolutely cannot tolerate an inconsistent system. Since we can't allow an inconsistent system, any system that we build *must* be incomplete. Saying that a system is incomplete means that there are true statements that can't be proven to be true.

Incompleteness spelled disaster for the Principia system—the entire point of the Principia effort was to make every true statement provable within a single formal system.

What does this have to do with Alan Turing and the halting problem?

Explaining the proof of incompleteness is difficult. There are entire books dedicated to the subject. We don't have the time or space to go through that! Fortunately, there's an easy way out in Alan Turing's proof of the so-called halting problem, which turns out to be a simpler version of nearly the same thing.

As we saw with Prolog in 14, *Programming with Logic*, on page 103, a logical inference system is a kind of *computing system*, because the process of searching for a proof in a logic *is* a computation. That means that if the Principia were capable of working, then for any statement in a logic, a search for a proof would eventually produce a proof that either the statement was true or that it was false. There would always be an answer, and that answer would always eventually be generated by a program.

So the question "Is there a proof that statement *S* is either true or false?" is really the same thing as the question "Will a program *P* ever finish with an answer?"

Gödel proved the logical side of that question: he proved that there are some statements that aren't false, but that you can't *prove* are true. Turing proved the equivalent statement that there are programs for which you can't tell whether they will ever finish, or *halt*.

They are, in a deep sense, the same thing. But understanding the proof of the halting problem is very easy compared to understanding Gödel's proof of incompleteness.

To Halt or Not To Halt?

To begin, we need to define what a computing device is. In formal mathematical terms, we don't care how it works; all we care about is what it can do in abstract terms. So we define something called an *effective computing device* or an *effective computing system*, abbreviated ECS. An effective computing device is any Turing-equivalent computing device: it could be a Turing machine or a λ calculus evaluator or a Brainf*** interpreter or the CPU in your mobile phone. I'm being deliberately vague here because we don't care what kind of machine it is. What we want to show is that *for any possible computing device*, it won't be able to tell correctly whether programs halt.

An ECS is modeled as a two-parameter function:

$$C : N \times N \rightarrow N$$

The first parameter is an encoding of a program as a natural number; the second parameter is the input to the program. It's also encoded as a natural number. Encoding it that way might seem limiting, but it really isn't because we can encode any finite data structure as a natural number. If the program halts, then the return value of the function is the result of the program. If the program doesn't halt, then the function doesn't return anything. In that case, we say that the pair consisting of the program and its input aren't in the domain of the ECS.

So if you wanted to describe running the program f on the input 7, you'd write that as $C(f, 7)$. And finally, the way that we would write that a program p doesn't halt for input i is $C(p, i) = \_$.

Now that we've got our basic effective computing device, we need to equip it to handle more than two parameters before we can use it to formulate the halting problem. After all, a halting oracle is a program that takes two inputs: another program and the input to that program. The easiest way to do that is to use a *pairing function*, a one-to-one function from an ordered pair to an integer.

There are lots of possible pairing functions. For example, you could convert both numbers to binary, left-pad the smaller of the two until they're of equal length, and then interleave their bits. Given (9, 3), you convert 9 to 1001, and 3 to 11; then you pad 3 to 0011 and interleave them to give you 10001011, or 139. It doesn't matter exactly which pairing function we use: what matters is that we know it's possible to choose *some* pairing function that lets us combine multiple parameters into a single number. We'll write our pairing function as $pair(x, y)$.

With the help of the pairing function, we can now express the halting problem. The question is, does there exist a program O, called a *halting oracle*, such that

$$\forall p, \forall i : C(O, pair(p, i)) = \begin{cases} 0 \text{ if } \phi(p, i) = \bot \\ 1 \text{ if } \phi(p, i) \neq \bot \end{cases}$$

In English: Does there exist a program *O* such that for all pairs of program *p* and inputs *i*, the oracle returns *1* if *C(pair(p, i))* halts, and 0 if it doesn't? Or less formally, can I write a program that tells whether or not other programs will eventually finish running and stop?

Now for the proof. Suppose that we *do* have a halting oracle, *O*. That means that for any program *p* and input *i*, *C(O, pair(p, i)) = 0* if and only if *C(pair(p, i))* = _.

What if we can devise a program p_d and input *i* where $C(p_d, i)$ halts, but *C(O, pair(p, i)) = 0*? If we can, then that will mean the halting oracle failed, which in turn will show that we *cannot* always determine whether a program will eventually halt.

So we're going to build that program. We'll call it *the deceiver*. The deceiver looks at what the halting oracle would predict that it would do, and then it does the opposite.

```
def deceiver(oracle) {
  if oracle(deceiver, pair(oracle, deceiver)) == 1 then
    loop forever
  else
    halt
}
```

Simple, right? Almost. It's not quite as easy as it might seem. You see, the problem is that the deceiver needs to be able to pass itself to the oracle. But how can it do that? A program can't pass *itself* into the oracle.

Why not? Because we're working with the program represented as a number. If the program contained a copy of itself, then it would have to be *larger than itself*. Which is, of course, impossible.

As an aside, there are a variety of tricks to work around this. One of the classic ones is based on the fact that for any given program *p*, there are an *infinite* number of versions of the same program with different numeric representations. Using that property, you can embed a program *d2* into a deceiver *d*. But there are a few other tricks involved in getting it right. It's not simple, and even Alan Turing screwed it up in the first published version of his proof!

Fortunately, there's a nice workaround. What we care about is whether there is any *combination* of program and input such that *O* will incorrectly predict the halting status. So we'll just turn the deceiver into a parameter that it can pass to itself. That is, the deceiver is this:

```
def deceiver(input) {
  (oracle, program) =  unpair(input)
  if oracle(program, input):
    while(True): continue
  else:
    halt
}
```

Then we'll be interested in the case where the value of the *program* parameter is the numeric form of the deceiver itself.

Now, when *input = pair(O, deceiver)*, *O* will make the wrong prediction about what *deceiver* will do. That means that once again we're back where we were in the simpler version of the proof. A halting oracle is a program that, given any pair of program and input, will correctly determine whether that program will halt on that input. We're able to construct a program and input pair where the oracle *doesn't* make the right prediction, and therefore it *isn't* a halting oracle.

This proof shows that *any* time anyone claims to have a halting oracle, they're wrong. And you don't need to take it on faith: this proof shows you how to build a specific example where the oracle will be wrong.

The halting problem seems like a simple one. Given a computer program, you want to know if it will ever finish running. Thanks to Turing, we know that's a question you can't answer. But the problem goes beyond the concerns of computer professionals because computation is so central to the field of mathematics. Long before anyone had actually thought deeply about computation, the limits of computation were setting the limits of mathematics. The fact that we can't know whether a program will ever stop means that there are problems in computing that we can't ever solve, and, more crucially, that mathematics itself can't ever solve.

If you're interested in learning more about Gödel and the incompleteness theorems, I highly recommend you check out two books. The first, *Gödel, Escher, Bach: An Eternal Golden*

Braid [Hof99], is my all-time favorite nonfiction book, and it does a beautiful job of actually walking you through the steps of Gödel's proof in an engaging, fun, informal way. If you prefer a more formal and mathematical approach, I recommend *Gödel's Proof [NN08]*, a brilliant presentation.

Bibliography

[CGP99] Edmund M. Clarke Jr, Orna Grumberg, and Doron A. Peled. *Model Checking*. MIT Press, Cambridge, MA, 1999.

[CM03] William F. Clocksin and Christopher S. Mellish. *Programming in Prolog: Using the ISO Standard*. Springer, New York, NY, USA, Fifth, 2003.

[Cha02] Gregory J. Chaitin. *The Limits of Mathematics: A Course on Information Theory and the Limits of Formal Reasoning*. Springer, New York, NY, USA, 2002.

[FFB95] Daniel P. Friedman, Matthias Felleisen, and Duane Bibby. *The Little Schemer*. MIT Press, Cambridge, MA, Fourth, 1995.

[Hod98] Wilfrid Hodges. An editor recalls some hopeless papers. *The Bulletin of Symbolic Logic*. 4[1], 1998, March.

[Hof99] Douglas R. Hofstadter. *Gödel, Escher, Bach: An Eternal Golden Braid*. Basic Books, New York, NY, USA, 20th Anniv, 1999.

[Lep00] Ernest Lepore. *Meaning and Argument: An Introduction to Logic Through Language*. Wiley-Blackwell, Hoboken, NJ, 2000.

[MHMT97] Robin Milner, Robert Harper, David MacQueen, and Mads Tofte. *The Definition of Standard ML - Revised*. MIT Press, Cambridge, MA, Revised, 1997.

[NN08] Ernest Nagel and James Newman. *Gödel's Proof*. NYU Press, New York, NY, 2008.

[O'K09] Richard O'Keefe. *The Craft of Prolog (Logic Programming)*. MIT Press, Cambridge, MA, 2009.

[Pie02] Benjamin C. Pierce. *Types and Programming Languages*. MIT Press, Cambridge, MA, 2002.

[Ray96] Eric S. Raymond. *The New Hacker's Dictionary*. MIT Press, Cambridge, MA, Third, 1996.

[Wol02] Stephen Wolfram. *A New Kind of Science*. Wolfram Media, Champaign, IL, 2002.

Put the "Fun" in Functional

Elixir puts the "fun" back into functional programming, on top of the robust, battle-tested, industrial-strength environment of Erlang.

Programming Elixir

You want to explore functional programming, but are put off by the academic feel (tell me about monads just one more time). You know you need concurrent applications, but also know these are almost impossible to get right. Meet Elixir, a functional, concurrent language built on the rock-solid Erlang VM. Elixir's pragmatic syntax and built-in support for metaprogramming will make you productive and keep you interested for the long haul. This book is *the* introduction to Elixir for experienced programmers.

Dave Thomas
(240 pages) ISBN: 9781937785581. $36
http://pragprog.com/book/elixir

Programming Erlang (2nd edition)

A multi-user game, web site, cloud application, or networked database can have thousands of users all interacting at the same time. You need a powerful, industrial-strength tool to handle the really hard problems inherent in parallel, concurrent environments. You need Erlang. In this second edition of the bestselling *Programming Erlang*, you'll learn how to write parallel programs that scale effortlessly on multicore systems.

Joe Armstrong
(510 pages) ISBN: 9781937785536. $42
http://pragprog.com/book/jaerlang2

Seven Databases, Seven Languages

There's so much new to learn with the latest crop of NoSQL databases. And instead of learning a language a year, how about seven?

Seven Databases in Seven Weeks

Data is getting bigger and more complex by the day, and so are your choices in handling it. From traditional RDBMS to newer NoSQL approaches, *Seven Databases in Seven Weeks* takes you on a tour of some of the hottest open source databases today. In the tradition of Bruce A. Tate's *Seven Languages in Seven Weeks*, this book goes beyond basic tutorials to explore the essential concepts at the core of each technology.

Eric Redmond and Jim R. Wilson
(354 pages) ISBN: 9781934356920. $35
http://pragprog.com/book/rwdata

Seven Languages in Seven Weeks

You should learn a programming language every year, as recommended by *The Pragmatic Programmer*. But if one per year is good, how about *Seven Languages in Seven Weeks*? In this book you'll get a hands-on tour of Clojure, Haskell, Io, Prolog, Scala, Erlang, and Ruby. Whether or not your favorite language is on that list, you'll broaden your perspective of programming by examining these languages side-by-side. You'll learn something new from each, and best of all, you'll learn how to learn a language quickly.

Bruce A. Tate
(330 pages) ISBN: 9781934356593. $34.95
http://pragprog.com/book/btlang

Tinker, Tailor, Solder, and DIY!

Get into the DIY spirit with Raspberry Pi or Arduino. Who knows what you'll build next...

Raspberry Pi

The Raspberry Pi is a $35, full-blown micro computer that runs Linux. Use its video, audio, network, and digital I/O to create media centers, web servers, interfaces to external hardware—you name it. And this book gives you everything you need to get started.

Maik Schmidt
(149 pages) ISBN: 9781937785048. $17
http://pragprog.com/book/msraspi

Arduino

Arduino is an open source platform that makes DIY electronics projects easier than ever. Even if you have no electronics experience, you'll be creating your first gadgets within a few minutes. Step-by-step instructions show you how to build a universal remote, a motion-sensing game controller, and many other fun, useful projects. This book has now been updated for Arduino 1.0, with revised code, examples, and screenshots throughout. We've changed all the book's examples and added new examples showing how to use the Arduino IDE's new features.

Maik Schmidt
(272 pages) ISBN: 9781934356661. $35
http://pragprog.com/book/msard

The Pragmatic Bookshelf

The Pragmatic Bookshelf features books written by developers for developers. The titles continue the well-known Pragmatic Programmer style and continue to garner awards and rave reviews. As development gets more and more difficult, the Pragmatic Programmers will be there with more titles and products to help you stay on top of your game.

Visit Us Online

This Book's Home Page

http://pragprog.com/book/mcmath

Source code from this book, errata, and other resources. Come give us feedback, too!

Register for Updates

http://pragprog.com/updates

Be notified when updates and new books become available.

Join the Community

http://pragprog.com/community

Read our weblogs, join our online discussions, participate in our mailing list, interact with our wiki, and benefit from the experience of other Pragmatic Programmers.

New and Noteworthy

http://pragprog.com/news

Check out the latest pragmatic developments, new titles and other offerings.

Save on the eBook

Save on the eBook versions of this title. Owning the paper version of this book entitles you to purchase the electronic versions at a terrific discount.

PDFs are great for carrying around on your laptop—they are hyperlinked, have color, and are fully searchable. Most titles are also available for the iPhone and iPod touch, Amazon Kindle, and other popular e-book readers.

Buy now at *http://pragprog.com/coupon*

Contact Us

Online Orders:	*http://pragprog.com/catalog*
Customer Service:	*support@pragprog.com*
International Rights:	*translations@pragprog.com*
Academic Use:	*academic@pragprog.com*
Write for Us:	*http://pragprog.com/write-for-us*
Or Call:	+1 800-699-7764

CPSIA information can be obtained at www.ICGtesting.com
Printed in the USA
LVOW01s1824180713

343550LV00002B/2/P

9 781937 785338